CRITICAL THINKING & LOGICAL REASONING WORKBOOK-3

GIFT OF LOGIC™ SERIES

An Essential Resource for Everyone

Boost Your Thinking Skills

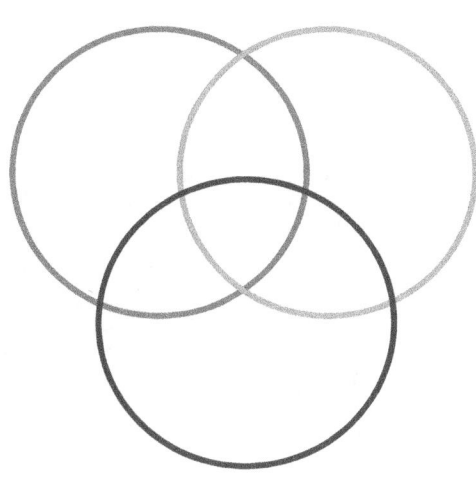

Verbal Reasoning

Analytical Reasoning

Pictorial Reasoning

THIRD EDITION

| FOR GRADES 3-5 | STUDENTS, TEACHERS, AND PARENTS |

Ranga Raghuram **GIFT OF LOGIC™**

 **Gift Of Logic, Inc**

http://www.giftoflogic.com
sales@giftoflogic.com

Critical Thinking and Logical Reasoning Workbook-3
ISBN-13: 978-1494832261
ISBN-10: 1494832267

Third Edition
1-2014

Copyright © 2009 Gift Of Logic, Inc. All rights reserved. No part of this publication may be reproduced, stored in a retrieval system, transmitted in any form or by any means, electronic, mechanical, photocopying, recording or otherwise, without the written permission of the publisher.

License: This book is licensed for use by one person only. Use of this book in a group setting (classroom, workshop, etc) without the written permission of the publisher is prohibited. Unauthorized duplication is strictly prohibited by law. Contact the publisher at sales@giftoflogic.com for classroom/school/group licensing.

GIFT OF LOGIC™
CRITICAL THINKING & LOGICAL REASONING CURRICULUM
12 WORKBOOKS TO BOOST YOUR THINKING SKILLS

For Kindergarten, Grade 1, and Grade 2

Workbook# 0

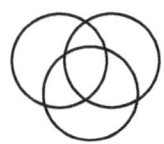

Verbal Reasoning	Finding the truth, Inferencing, Analogies, Synonyms and Antonyms, Agree/Disagree
Analytic Reasoning	Memory drill, Decision making, Positioning, Sudoku
Pictorial Reasoning	Connect the dots, Mazes, Picture Sequence, Spot the difference, etc

Workbook# 1

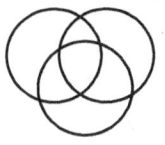

Verbal Reasoning	Finding the truth, Inferencing, Analogies, Synonyms and Antonyms, Agree/Disagree
Analytic Reasoning	Sorting, Positioning, Picking, Assorted problems, Numeric and Alphabetic Sudoku
Pictorial Reasoning	Picture Sequence, Spot the difference, Odd picture

Workbook# 2

Verbal Reasoning	Finding the truth, Classification, Direct and Inverse relationship, Inferencing, Analogies, Agree/Disagree
Analytic Reasoning	Sequencing, Scheduling, Strategy, Picking, etc
Pictorial Reasoning	Picture Analogy, Odd picture, Pattern matching, etc

For Grade 3, Grade 4, and Grade 5

Workbook# 3

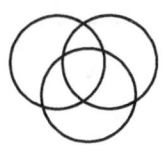

Verbal Reasoning	Not, And, Or, If .. then, Conditional inferencing, Unconditional inferencing, Symbolic Logic
Analytic Reasoning	Lists, Sequencing, Grouping, Venn Diagrams, Graph logic, Number logic, Letter logic, Sudoku
Pictorial Reasoning	Picture sequence, Picture analogy, Odd picture, Picture difference, Pattern matching

Workbook# 4

Verbal Reasoning	Contradiction, Converse, Inverse, Contrapositive, Conditional inferencing, Symbolic Logic
Analytic Reasoning	Scheduling, Looping, FIFO, LIFO, Correlation, Venn Diagram, Graph logic, Number logic, Sudoku, etc
Pictorial Reasoning	Picture sequence, Picture analogy, Odd picture, Picture difference, Pattern matching

Workbook# 5

Verbal Reasoning	Biconditional, Categorical inferencing, Cause and Effect, Symbolic Logic, Agree/Disagree, Word and Sentence analogy
Analytic Reasoning	Correlation, Grouping, Venn Diagrams, Graph logic, Number logic, Letter logic, Sudoku, etc
Pictorial Reasoning	Picture sequence, Picture analogy, Odd picture, Picture difference, Pattern matching

********* Essential resource for everyone *********
*http://www.giftoflogic.com *sales@giftoflogic.com

GIFT OF LOGIC™
CRITICAL THINKING & LOGICAL REASONING CURRICULUM
12 WORKBOOKS TO BOOST YOUR THINKING SKILLS

For Grades 6-12, College/University Students, Adults

Primer

Prereq

Verbal Reasoning	Logical Operators, Conditional, Categorical and Causal reasoning, Validity, Fallacies, Symbolic Logic
Analytic Reasoning	Positioning, Grouping, Sudoku
Pictorial Reasoning	Pattern perception, Figure formation, Paper folding and cutting, Figure matrix, Rule detection

Workbook# 6

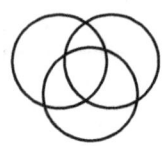

Verbal Reasoning	Arguments-Main point, Must be true, Cannot be true
Analytic Reasoning	Positioning, Grouping, Sudoku
Pictorial Reasoning	Pattern perception, Figure formation, Paper folding and cutting, Figure matrix, Rule detection

Workbook# 7

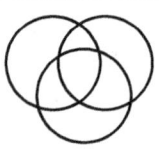

Verbal Reasoning	Arguments-Strengthening, Weakening
Analytic Reasoning	Positioning, Grouping, Sudoku
Pictorial Reasoning	Pattern perception, Figure formation, Paper folding and cutting, Figure matrix, Rule detection

Workbook# 8

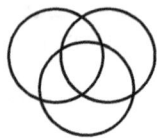

Verbal Reasoning	Arguments - Controversy, Paradox
Analytic Reasoning	Positioning, Grouping, Sudoku
Pictorial Reasoning	Pattern perception, Figure formation, Paper folding and cutting, Figure matrix, Rule detection

Workbook# 9

Verbal Reasoning	Arguments- Assumptions, Reasoning strategy
Analytic Reasoning	Positioning, Grouping, Sudoku
Pictorial Reasoning	Pattern perception, Figure formation, Paper folding and cutting, Figure matrix, Rule detection

Workbook# 10

Verbal Reasoning	Arguments-Flawed reasoning, Analogous reasoning
Analytic Reasoning	Positioning, Grouping, Sudoku
Pictorial Reasoning	Pattern perception, Figure formation, Paper folding and cutting, Figure matrix, Rule detection

********* Essential resource for everyone *********
Get the GIFT OF LOGIC™ today !
*http://www.giftoflogic.com *sales@giftoflogic.com

Dear Reader:

Your decision to purchase this book is commendable. You now have in your hands, a comprehensive, easy-to-read book in Critical thinking and Logical reasoning that will introduce you to three different areas of thinking and reasoning - Verbal, Analytical and Pictorial. Solving problems in Verbal Reasoning is important to develop a critical mind. Solving problems in Analytic Reasoning is important to develop a flexible and resourceful mind. Solving problems in Pictorial Reasoning is important to develop a visually alert mind.

This book is presented in a workbook format to help you progress quickly. Parents and teachers are urged to complete the exercises ahead of the student and assist them whenever necessary with the help of detailed answers provided at the end of the book. This book can be used as a supplementary resource in the regular class room or it can be used during winter and summer vacations. College/ University students, working professionals and retired individuals will also find the Gift Of Logic(tm) Series very useful in enhancing their problem solving abilities, confidence and general intellect.

Critical thinking and Logical reasoning must be practiced consistently to develop strong cognitive skills. After completing the exercises in this book, continue to read the other books in this series to get familiar with different types of Logical reasoning problems.

This workbook is one in a series of twelve workbooks. Please refer to the brochure before this page for a brief description of each workbook. Visit the website http://www.giftoflogic.com for more information.

Happy thinking and reasoning!

TABLE OF CONTENTS

Verbal Reasoning

Logical Negation (NOT)...9

Logical Conjunction (AND)...12

Logical Disjunction (OR)...15
    Inclusive OR
    Exclusive OR

Conditional Statements
    If..then..20
    Unless...25
    Except...28

Inferencing
    conditional - must be true...30
    conditional - cannot be true..33

Inferencing
    unconditional - must be true..34
    unconditional - cannot be true...38

Agree...42

Disagree..46

TABLE OF CONTENTS

Analytic Reasoning

List processing..49
Sequencing..57
Grouping..61
Venn Diagrams..64
Graph Logic..69
Number Logic...76

Letter Logic..79

Sudoku...82

Pictorial Reasoning

Picture Sequence..88
Picture Analogy..91
Odd Picture..94
Picture Difference..96
Pattern Matching..99

Answers

Verbal..101
Analytic...128
Pictorial..158
Certificate of Completion

Name ——————————————— Date———————————————

VERBAL REASONING

Name _____   Date_____

Logical Negation using NOT

A statement that is true or false can be negated using the word "NOT".

Negation is represented by the symbol ~ or !

Negation of a statement P is symbolically represented by ~P or !P

Negation of a true statement is a false statement. Negation of a false statement is a true statement. Another way to express the negation of a statement is by using a statement that begins with "It is not the case that".

Example:

P	~P (Negation of P)
Rome is the capital of Italy (true)	Rome is not the capital of Italy (false)
Ireland is in Africa (false)	Ireland is not in Africa (true)

In the above table, note that if P is true, then ~P is false and if P is false, then ~P is true. The following table is called the truth table for NOT.

Truth table for NOT	
P	~P
true	false
false	true

Example:
Negate the following statement.
    Statement: All apples are red.
    Negation 1:   All apples are not red.
    Negation 2:   It is not the case that all apples are red.

Verbal Reasoning
© Gift Of Logic, Inc * Copying prohibited

Name _____   Date _____

Logical Negation using NOT

Negate the following statements and answer the question.

1 Statement: World War II happened in 1990.
Negation:
Negation is   A) True   B) False

2 Airplanes travel in the ocean.
Negation:
Negation is   A) True   B) False

3 Statement: Mercury is the planet closest to the Sun.
Negation:
Negation is   A) True   B) False

4 Statement: Wild animals are not found in the zoo.
Negation:
Negation is   A) True   B) False

5 Statement: Computers can not play chess.
Negation:
Negation is   A) True   B) False

6 Statement: All doctors are women.
Negation:
Negation is   A) True   B) False

Verbal Reasoning   Answers-101
© Gift Of Logic, Inc * Copying prohibited

Name _____ Date _____

Symbolic Representation of NOT (~)

A negation can be represented in short form using the NOT symbol ~. Doing so helps to represent the facts in a concise, logical format that can be used for making inferences.

In the statements below, negate the underlined portion using the ~ symbol.

Example:
    Statement: World War II <u>happened in 1990.</u>
    Symbolic Negation: ~happen in 1990

---

1   Airplanes <u>travel in the ocean.</u>
    Symbolic Negation:

---

2   Statement: Mercury is the planet <u>closest</u> to the Sun.
    Symbolic Negation:

---

3   Statement: Wild animals are <u>not found in the zoo</u>.
    Symbolic Negation:

---

4   Statement: Computers <u>cannot play chess.</u>
    Symbolic Negation:

---

5   Statement: <u>All doctors are women</u>.
    Symbolic Negation:

Verbal Reasoning
© Gift Of Logic, Inc * Copying prohibited

Name _____   Date _____

Logical Conjunction using AND

A connection of two statements by the word "AND" is called a conjunction. Conjunction is represented by the symbol &. Conjunction of two statements P and Q is represented by P&Q. Statements P and Q are called conjuncts.

    Conjunction: Rick is tall <u>and</u> Rick is smart.
    Symbolic representation: Rick is tall & Rick is smart.
    Conjuncts: Rick is tall, Rick is smart.

Both the conjuncts must be true for the conjunction to be true. If one of the conjuncts is false or both the conjuncts are false, then the conjunction is false. Examples are shown in the table below.

P	Q	P&Q
Rick is tall True	Rick is smart True	Rick is tall & smart True
Rick is tall True	Rick is smart False	Rick is tall & smart False
Rick is tall False	Rick is smart True	Rick is tall & smart False
Rick is tall False	Rick is smart False	Ricks is tall & smart False

The following table is called the truth table for AND. The table shows what the conjunction (P&Q) would be for different values of conjuncts P and Q.

Truth table for AND		
P	Q	P & Q
true	true	true
true	false	false
false	true	false
false	false	false

Verbal Reasoning

Name _____ Date _____

Logical Conjunction using AND

In the following exercises, find the truth of the conjunction.

1 Conjunction: Some apples are green and some apples are blue.

Conjunction is A) True   B) False

---

2 Conjunction: Human beings can jump and fly.

Conjunction is A) True   B) False

---

3 Conjunction: Cotton can catch fire and cotton is not soft.

Conjunction is A) True   B) False

---

4 Conjunction: Some buildings are tall and some buildings are short.

Conjunction is A) True   B) False

---

5 Conjunction: A pen and a guitar are musical instruments.

Conjunction is A) True   B) False

---

6 Conjunction: A car has two wheels and a bike has no wheels.

Conjunction is A) True   B) False

Verbal Reasoning
© Gift Of Logic, Inc * Copying prohibited

Name _____ Date _____

Symbolic Representation of Conjunction (&)

A conjunction can be represented in short form using the AND symbol &.

Represent the following conjunctions in symbolic form using &.

Example:
   Conjunction: Rick is tall and smart.
   Symbolic Representation: tall & smart

1. Conjunction: A car has four wheels and a bike has no wheels.
   Symbolic:

2. Conjunction: Human beings can jump and fly.
   Symbolic:

3. Conjunction: Cotton can catch fire and cotton is not soft.
   Symbolic:

4. Conjunction: She ranked first in singing and third in swimming.
   Symbolic:

5. Conjunction: A pen and a guitar are musical instruments.
   Symbolic:

Verbal Reasoning   Answers-104
© Gift Of Logic, Inc * Copying prohibited

Name _____ Date _____

Logical Disjunction using OR

Two statements connected by the word "OR" is called a disjunction.
Disjunction is represented by the symbol ∥.
Disjunction of two statements P and Q is represented as P∥Q.
Statements P and Q are called disjuncts.

Think of each line in ∥ as representing one disjunct.

Just one disjunct has to be true for the entire disjunction to be true. If both disjuncts are false, then the entire disjunction is false. Examples are shown in the table below.

P	Q	P∥Q
Rick is tall True	Rick is smart True	Rick is tall or smart True
Rick is tall True	Rick is smart False	Rick is tall or smart True
Rick is tall False	Rick is smart True	Rick is tall or smart True
Rick is tall False	Rick is smart False	Rick is tall or smart False

Truth table for OR		
P	Q	P∥Q
true	true	true
true	false	true
false	true	true
false	false	false

The disjunction "Rick is tall or smart" can be written in symbolic form as tall∥smart. Since Rick can be tall or smart or both, the symbol ∥ includes both possibilities. So, this type of "or" is called an "Inclusive OR". Another type of "OR" is called the "Exclusive OR".

Verbal Reasoning
© Gift Of Logic, Inc * Copying prohibited

Name _____ Date _____

Logical Disjunction   Inclusive OR - Exclusive OR

The logical disjunction "OR" can be used inclusively or exclusively. You have to determine yourself if it is inclusive or exclusive.

INCLUSIVE OR ∥

In statements that have an inclusive OR, both the disjuncts can be true.

Example: Let us eat breakfast or lunch together.

In this example, it is logically possible for two people to eat breakfast together, or lunch together, or eat both together. So, the "OR" in the statement is used in an inclusive (both) context. In symbolic form, this statement can be represented as, breakfast ∥ lunch.

EXCLUSIVE OR ╫

In statements where the "OR" is exclusive, only one disjunct can be true. Both the disjuncts cannot be true at the same time.

Example: John is currently upstairs or downstairs. In this example, John cannot be in two places at the same time. So, he is either upstairs or downstairs, but not both. The "OR" in the statement is used in an exclusive (but not both) context. In symbolic form, this can be represented as upstairs ╫ downstairs. Think of the horizontal line as representing the fact that both options cannot be true.

Sometimes, exclusive OR is stated clearly using " but not both".
Example: Rick will play the piano or the guitar, but not both.
In this example, it is very clear that Rick will play only one instrument. The "but not both" in the statement makes this clear. The symbolic form for this statement is piano ╫ guitar. We still do not know what he will play, but we do know that he will play only one instrument.

Verbal Reasoning

Name _____   Date _____

Logical Disjunction   Inclusive OR - Exclusive OR

In questions 1-5, read the disjunction statement and find out if it is an inclusive OR or an exclusive OR.

1 A tiger can live in a forest or a zoo.

The "OR" in the statement is an
    A) inclusive OR   B) exclusive OR

Tigers can live in
    A) a forest only   B) a zoo only   C) a forest, a zoo, or both
    D) either a forest or a zoo, but not both

---

2 John can be in team A or team B or both.

The "OR" in the statement is an
    A) inclusive OR   B) exclusive OR

John can belong to which of the following teams?
    A) A only   B) B only   C) A and B

---

3 Lisa can be in team A or team B, but not both.

The "OR" in the statement is an
    A) inclusive OR   B) exclusive OR

Lisa can belong to which of the following teams?
    A) A   B) B   C) A and B

Verbal Reasoning
© Gift Of Logic, Inc * Copying prohibited

Name _____   Date _____

| Logical Disjunction   Inclusive OR - Exclusive OR |

4 Mary can sit in the first chair or the fourth chair now.

The "OR" in the statement is an
       A) inclusive OR   B) exclusive OR

Mary can sit in which of the following chairs?
    A) first only   B) first or fourth   C) first and fourth

---

5 Tom can eat an apple or an orange, but not both.

The "OR" in the statement is an
       A) inclusive OR   B) exclusive OR

At most, how many fruits can Tom eat?
    A) 0   B) 1

Verbal Reasoning
© Gift Of Logic, Inc * Copying prohibited

Name _____ Date _____

| **Symbolic Representation of Logical Disjunction - (∥) (⫻)** |

Represent the following statements by using the symbolic notation for OR.
Use the symbols ∥ for inclusive OR and ⫻ for exclusive OR.

Example:

    Disjunction: A tiger can live in a forest or a zoo.
    Symbolic: forest ∥ zoo

---

1   Disjunction: John can be in team A or team B or both.
    Symbolic:

---

2   Disjunction: Lisa can be in team A or team B, but not both.
    Symbolic:

---

3   Disjunction: Mary can sit in the first chair or the fourth chair.
    Symbolic:

---

4   Disjunction: Tom can eat an apple or an orange.
    Symbolic:

---

5   Disjunction: The tea is served either hot or cold.
    Symbolic:

Verbal Reasoning
© Gift Of Logic, Inc * Copying prohibited

Name _____ Date _____

CONDITIONAL STATEMENTS   If..then

You might have come across conditional statements like the ones below:

    If you behave well, then you can go to the park.
    If you eat apples, you will be healthy.

Conditional statements are normally expressed in a "if..then" format, but there are other ways of expressing conditions without using "if..then". We will learn about expressing conditions using words such as Unless and Except later in this book.

The word "then" in the "if..then" format is not necessary, but using it gives clarity to the condition. Note that the word "if" can appear in the middle of a conditional statement.

<u>Examples of conditional statements</u> (note the location of "if" and "then"):

    If Martha dances, then Mark will sing.
    Mark will sing if Martha dances.
    If the volcano erupts, ash will be thrown out.
    The landscape will be pretty if the flowers bloom.
    If the baby is hungry, it will cry.

In the statement "If P then Q", P is called the antecedent and Q is called the consequent. If the words "antecedent" and "consequent" are hard to remember, then think of these parts as the "if-part" and the "then-part". The antecedent is the "if-part". The consequent is the "then-part". If the antecedent is true, then the consequent will be true. In other words, if the antecedent happens, the consequent will definitely happen. This is an important concept to understand.

Name _____ Date _____

| Symbolic Representation of If P then Q (→) |

A conditional statement "If P then Q" can be symbolically represented using an arrow as follows: P→Q

P is the antecedent; Q is the consequent.

Conditional: If Martha dances, then Mark will sing.
Symbolic Representation: Martha dances→Mark will sing.

Represent the following conditional statements in symbolic form. Identify the antecedent and consequent as well.

---

1 Conditional: If Martha cries, then Mark will smile.

Symbolic:

Antecedent: Consequent:

---

2 Conditional: If the baby is hungry, it will cry.

Symbolic:

Antecedent: Consequent:

---

3 Conditional: Mark will sing if Martha dances.

Symbolic:

Antecedent: Consequent:

Verbal Reasoning   Answers-108
© Gift Of Logic, Inc * Copying prohibited

Name _____ Date _____

| CONDITIONAL STATEMENTS   If P then Q |

Follow the instructions in each question below and make up a few conditional statements yourself.

1 Write a conditional statement of the form "If P then Q"

Symbolic Representation:
What is the antecedent?
What is the consequent?

2 Write a conditional statement without using "then"

Symbolic Representation:
What is the antecedent?
What is the consequent?

3 Write a conditional statement with "if" in the middle of the statement.

Symbolic Representation:
What is the antecedent?
What is the consequent?

Verbal Reasoning

Name _____ Date _____

CONDITIONAL STATEMENTS- If P then Q using "Not"

Conditional statements <u>can</u> also use the word "not" in their antecedent or their consequent or both. Remember that "Not" is represented by ~, the symbol for negation.

"Not" in antecedent:
> If the tornado does not strike our city, we will be safe.
> ~tornado strike → safe

Mentally, read the above symbolic form as "if not tornado strike, then we will be safe". Even though, this statement is not grammatically correct, mentally, the symbolic form makes logical sense.

"Not" in the consequent:
> If you have homework, you must not watch a movie.
> have homework → ~watch movie

Mentally, read the above symbolic form as "if you have homework, then you must not watch movie".

"Not" in both antecedent and consequent.
> If you do not score a goal, you will not get a prize.
> ~score a goal → ~get a prize

Mentally, read the above symbolic form as "if you do not score a goal, then you will not get a prize".

Verbal Reasoning

Name _____ Date _____

| CONDITIONAL STATEMENTS   If P then Q using "Not" |

1 Write a conditional statement with "not" in its antecedent. Represent the statement in symbolic form.

Symbolic: _____

2 Write a conditional statement with "not" in its consequent. Represent the statement in symbolic form.

Symbolic: _____

3 Write a conditional statement with "not" in both its antecedent and consequent.

Symbolic:

Verbal Reasoning   Answers-109
© Gift Of Logic, Inc * Copying prohibited

CONDITIONAL STATEMENTS - Unless

Conditional statements can be expressed using the word "Unless". This word can be used in the beginning or in the middle of a sentence. Logically, "unless" is the same as "if not".

Examples:
    Unless you eat, we will not go to the movie.
    We will not go to the movie unless you eat.
    Unless you graduate you will not get a job.
    We will not be healthy unless we control air pollution.
    Unless we paint, our house will look dirty.
    Water will become scarce unless we stop deforestation.

Converting a conditional statement with the word "Unless" to a "if ..then" statement helps us to understand the condition clearly.

Unless in the beginning:
Conditional statement with Unless:	Unless P, Not Q
Conversion to if..then:	If Not P, then Not Q
Symbolic:	$\sim P \rightarrow \sim Q$

Unless in the middle:
Conditional statement with Unless:	Not Q, unless P
Conversion to if..then:	Not Q, if Not P
Symbolic:	$\sim P \rightarrow \sim Q$

Note that Unless P becomes a "If Not P" in the above two scenarios. Note the positions of P and Q in both cases.

Verbal Reasoning

Name _____ Date _____

CONDITIONAL STATEMENTS - Unless

The following examples show how conditional statements that use "Unless" are converted to "If..then" statements.

<u>"Unless" in the beginning of the conditional statement:</u>

Conditional: <u>Unless</u> you turn on the switch, the bulb will not glow.
If then: If you do not turn on the switch, the bulb will not glow.

Symbolic: ~turn on the switch → ~bulb will glow

Conditional: <u>Unless</u> you eat, we will not go to the movie.
If..then: If you do not eat, we will not go to the movie.
Symbolic: ~eat → ~go to movie

<u>"Unless" in the middle of the conditional statement:</u>

Conditional: We must not pollute <u>unless</u> we want to be sick.
If..then: We must not pollute if we do not want to be sick.
Symbolic: ~want to be sick → ~pollute

Conditional: We will not go to the zoo <u>unless</u> you finish your homework.
If..then: We will not go to the zoo if you do not finish your homework.
Symbolic: ~finish your homework → ~go to zoo

Verbal Reasoning
© Gift Of Logic, Inc * Copying prohibited

Name _____ Date _____

CONDITIONAL STATEMENTS - Unless

Convert the following conditional statements that use "Unless" into conditional statements and write their symbolic form.

---

1 Unless: Unless it stops raining, the shuttle will not take off.

If.. then:

Symbolic:

---

2 Unless: Unless everyone is silent, the show will not begin.

If.. then:

Symbolic:

3

Unless: Unless you exercise, you will not be healthy.

If.. then:

Symbolic:

---

4 Unless: You must not leave, unless I come back from shopping.

If.. then:

Symbolic:

Name _____ Date _____

CONDITIONAL STATEMENTS - Except

Conditions can be expressed using the word "Except". This word can be used in the beginning or in the middle of a sentence. These statements also will have words like must, should, necessary, etc. Logically, "except" is the same as "if not".

Examples:
    Except Rosie, everyone must sing twice.
    Everyone must play, except Rosie.

Converting conditional statements with the word "Except" to "If ..then" statements helps us to understand the condition clearly.

Except in the beginning:
Condition: Except P, Q
Conversion to If..then: If Not P, then Q
Symbolic: $\sim P \rightarrow Q$

Condition: Except Rosie, everyone must sing twice.
If..then: If you are not Rosie, you must sing twice.
Symbolic: $\sim Rosie \rightarrow sing\ twice$

Except in the middle:
Condition: Q, except P
Conversion to If..then: If Not P, then Q
Symbolic: $\sim P \rightarrow Q$
Except: Everyone must play, except Rosie.
If..then: If you are not Rosie, you must play.
Symbolic: $\sim Rosie \rightarrow play$

Verbal Reasoning
© Gift Of Logic, Inc * Copying prohibited

Name _____ Date _____

CONDITIONAL STATEMENTS - Except

Convert the following conditional statements that use "Except" into conditional statements and write their symbolic form.

1

Condition: Except on Wednesdays, you must wear a tie every day.
If..then:
Symbolic:

2

Condition: Except during winter, you must walk to school.
If..then:
Symbolic:

3

Condition: Everyone must practice at 6 PM except George.
If..then:
Symbolic:

4

Condition: The park must be closed for public except when it is summer.
If..then:
Symbolic:

5

Condition: Except on Mondays, the recreation center is open everyday.
If..then:
Symbolic:

Verbal Reasoning
© Gift Of Logic, Inc * Copying prohibited

Name _____   Date _____

| INFERENCING   conditional-must be true |

1 Brandi's mom said that if she goes to the circus, then she can see a clown. Brandi went to the circus.

If the above statements are true, then which one of the following statements must be true?
- A) She saw a clown.
- B) She did not see a clown.

---

2 Andy's grandma: "If you try hard, you will be able to climb Mount Everest". Andy tried hard.
Which one of the following can be inferred from the statements above?
- A) Andy was able to climb Mount Everest.
- B) Andy was not able to climb Mount Everest.

---

3 If the bell rings, the students will go home. The bell rang.

If the above statements are true, then which one of the following statements must be true?
- A) The students went to the park.
- B) The students went home.

---

4 Doctors say that people will not fall sick if they get flu shots. Bobby got the flu shot.
Which one of the following can be inferred from the statements above?
- A) People will not fall sick.
- B) Bobby will not fall sick.

Verbal Reasoning   Answers-112
© Gift Of Logic, Inc * Copying prohibited

Name _____   Date _____

| INFERENCING   conditional-must be true |

5 Advertisement: Introducing the new pureWhite soap. If you use it on your dirty white clothes, they will become clean within two minutes!

If the advertisement is true, then which of the following <u>must be true</u>?

　　A) Joyce used the pureWhite soap on her muddy white shirts and it became clean within three minutes.

　　B) Joyce used the pureWhite soap on her muddy white pants and it became clean within two minutes.

---

6 Banner in library: "If you read adventure books, you will be thrilled".
Mary and John read two adventure books.

If the banner's claim is true, then which of the following <u>must be true</u>?

 A) Mary will be thrilled.
 B) John will be bored.

---

7 Cat to Mouse: If you run away, I will not eat you. The mouse ran away.

If the above statements are true, then which one of the following statements <u>must be true</u>?

A) The cat ate the mouse.
B) The cat did not eat the mouse.

Verbal Reasoning

Name _____ Date _____

| INFERENCING   conditional-must be true |

8 Fox to Crow: If you sing a song, I will be happy. The crow sang a song.

If the above statements are true, then which one of the following statements <u>must be true</u>?

  A) The Crow was happy.
  B) The Fox was happy.

---

9 Policeman: If you cross the street when you see a 'do not walk' signal, you will be fined. Benjamin and Martha crossed the street when it said "do not walk".

Based on the above facts, answer the following true or false questions.

1) Benjamin was not fined.
   A) True   B) False

2) Martha was fined.
   A) True   B) False

Name _____   Date _____

INFERENCING   conditional-cannot be true

10 Notice at the botanical gardens: If it is Spring, the birds will come!

If the above information is true, which of the following cannot be true?
 A) The birds did not come even though it is Spring.
 B) As soon as Spring arrived, the birds also arrived.

---

11 If Gibson comes to the picnic, then Larry will come to the picnic.
If the above information is true, which of the following cannot be true?
 A) Gibson came the picnic and Larry came to the picnic.
 B) Gibson came to the picnic, but Larry did not.

---

12 Doctor to Kerry: If you take this medicine, you will get well.
If the Doctor is correct, then which one of the following cannot be true?
 A) Kerry did not get well in spite of taking the medicine.
 B) Kerry got well after he took the medicine.

---

13 Bookstore: If you pay for the book today, it will be available the next day.
If this information is true, which one of the following cannot be true?
 A) Tom paid on Tuesday and got the book on Thursday.
 B) Rick paid for the book on Wednesday and got the book on Thursday.

Name _____ Date _____

| INFERENCING   unconditional-must be true |

15
Sam hit the ball. The ball went outside the fence and smashed a window.

Which of the following conclusions can be drawn?
    A) The window is inside the fence.
    B) The window is outside the fence.

16
The water comes out of the shower head. The shower head is attached to the pipe on the wall.

Which of the following conclusions can be drawn?
    A) The water comes out of the pipe.
    B) The shower head is attached to the wall.

17
Airplanes carry passengers. The passengers carry bags with them.

Which of the following conclusions can be drawn?
    A) The bags carry the passengers.
    B) Airplanes carry the bags.

18
The ship is sailing. The passengers are in the ship.

Which of the following conclusions can be drawn?
    A) Passengers are sailing.
    B) Passengers are not sailing.

Verbal Reasoning   Answers-116

Name —————————— Date ——————————

INFERENCING unconditional-must be true

19 From wood, we can make paper. From paper, we can make notebook.

If the above statements are true, which of the following also must be true?

    A) From notebook, we can make wood.
    B) From wood, we can make notebook.

---

20 The train is going north. Tracy is sitting in the train facing south.

If the above statements are true, which of the following also must be true?

    A) Tracy is going south.
    B) Tracy is going north.

---

21 All birds are beautiful. Peacock is a bird.

If the above statements are true, which of the following also must be true?

    A) Peacock is not beautiful.
    B) Peacock is beautiful.

---

22 All the students in a class are reading. Nick is a student in the class.

If the above statements are true, which of the following must be true?

    A) Nick is writing.
    B) Nick is reading.

Verbal Reasoning   Answers-117
© Gift Of Logic, Inc * Copying prohibited

Name _____ Date _____

| INFERENCING   unconditional-must be true |

23 above/below

Floor P is below floor Q. Floor R is below floor P.
If the above statements are true, which of the following also must be true?

    A) Floor R is above floor Q.
    B) Floor R is below floor Q.

---

24 before

There will be a test on Logic before lunch.
If the above statement is true, which of the following also must be true?

    A) When the test begins, lunch would have been taken.
    B) When the lunch begins, the test would have been taken.

---

25 after

During a field trip, Amy's class will visit a farmland. Then they will go to the Space Center.
If the above statements are true, which of the following also must be true?

    A) Amy's class will see tractors before they see rockets.
    B) Amy's class will see tractors after they see rockets.

---

26 limits

Jim is driving his car at 55 miles per hour and Jason is driving his car at 45 miles per hour on a highway which has a speed limit of 50 miles per hour. If the above statement is true, which of the following also must be true?

    A) Both Jim and Jason are driving above the speed limit.
    B) Only Jim is driving above the speed limit.

Name _____ Date _____

| INFERENCING   unconditional-must be true |

27 front/behind
A rabbit, a fox and a lion are standing in a line. The rabbit is standing in front of the fox. The fox is standing in front of the lion.
If the above statements are true, which of the following also must be true?
    A) The lion is standing behind the rabbit.
    B) The rabbit is standing behind the lion.

28 branching
A tree has several branches. A branch has several leaves.
If the above statements are true, which of the following also must be true?
    A) A leaf has several trees.
    B) A tree has several leaves.

29 exception
Everyone in Tim's class is at most three feet tall. John is the only exception.
If the above statements are true, which of the following also must be true?
    A) John is three feet tall.
    B) John is more than three feet tall.

30 except
All stores are open this Sunday except the Bond bakery.
If the above statement is true, which of the following also must be true? A) Nothing can be purchased at the Bond bakery.
    B) Nothing can be purchased at the Bond bakery this Sunday.

Verbal Reasoning

Name _____   Date_____

INFERENCING   unconditional-cannot be true

31 same
Two trains are going in the same direction. One train is going North.
If the above statements are true, which of the following statements cannot be true?
    A) The other train is going south.
    B) The other train is going north.

32 different
The two flowers in the flower vase are of different colors. One of the flowers is red.
If the above statements are true, which of the following statements cannot be true?
    A) The other flower is blue in color.
    B) The other flower is red in color.

33 not
A leader's job is not an easy job. Jack is the leader of his class.
If the above statements are true, which of the following statements cannot be true?
    A) Jack has an easy job.
    B) Jack has a tough job.

34 not
The group of people in this room are not boys. A person in this group is wearing a mask. If the above statements are true, which of the following statements cannot be true?
    A) The person wearing a mask is a boy.
    B) The person wearing a mask is a girl.

Verbal Reasoning   Answers-120

Name ——————————— Date ———————————

| INFERENCING   unconditional-cannot be true |

35 same
It does not matter whether you travel in flight# 33 or flight# 43. It will take you the same time to go to London in either of them.

If the above statements are true, which of the following statements cannot be true?

    A) Flight# 33 will land before flight# 43.
    B) Flight# 43 takes fifty minutes, whereas flight# 33 takes forty minutes to reach London.

36 one/all
One end of a battery is positive and the other end is negative. All batteries are made this way.
If the above statements are true, which of the following statements cannot be true?

 A) Both ends of batteries made by Bright Industries are positive.
 B) Both ends of batteries made by Bright Industries are of different types.

37 order
Several students stood in a line to the see the doctor. They will be checked by the doctor who will attend to them one by one starting with the first student in the line.
If the above statements are true, which of the following statements cannot be true?

 A) The third student was seen by the doctor before the fifth student.
 B) The fourth student was seen by the doctor before the second student.

Verbal Reasoning   Answers-121
© Gift Of Logic, Inc * Copying prohibited

Name _____   Date _____

INFERENCING   unconditional-cannot be true

38 limits
Jack can lift up to 100 pounds. Jim can lift up to 200 pounds.

If the above statements are true, which of the following statements cannot be true?
   A) Jim can lift whatever Jack can.
   B) Jack can lift whatever Jim can.

39 compare
Temperatures are warmer in summer when compared to other seasons. Water evaporates more during warmer months.

If the above statements are true, which of the following statements cannot be true?
   A) Water does not evaporate more during summer.
   B) Water evaporates more during summer.

40 not vice versa
The candy box can be placed inside the popcorn box, but not vice versa.

If the above statement is true, which of the following statements cannot be true?
   A) Popcorn box can be placed inside the candy box.
   B) Popcorn box cannot be placed inside the candy box.

Verbal Reasoning   Answers-122
© Gift Of Logic, Inc * Copying prohibited

| INFERENCING   unconditional-cannot be true |

41 vice versa

Jack went to Jill's house and vice versa.
If the above statement is true, which of the following statements cannot be true?

    A) Jill went to Jack's house.
    B) Jill did not go to Jack's house.

---

42 swap

Brendan was standing third in line and Barak was standing fifth in line before they decided to switch their positions.

If the above statements are true, which of the following statements cannot be true after the switch?

    A) Brendan is ahead of Barak in the line.
    B) Barak is ahead of Brendan in the line.

Name _____   Date _____

AGREE

1 In questions 1-4, someone will be expressing their opinion to you on some subject. You need to reply by agreeing with whatever the person says, but using your own words.

Nurse: Eating food with too much sugar is not good for your health.

Your reply:

2

Car driver: While driving a car, we must keep a safe distance from the car in front of us.

Your reply:

Verbal Reasoning

Name _____ Date _____

AGREE

3

Civil Engineer: Bridges must have a very strong foundation.

Your reply:

4

Policeman: Wearing a helmet is very important when you ride your bike.

Your reply:

Name _____   Date _____

AGREE

5 In questions 5-8, read the conversation carefully and answer the questions.

Fox: The lion is majestic. So, it is the king of the jungle.

Pig: The lion is powerful. So, it is the king of the jungle.

What do the Fox and the Pig agree about?
    A) that the lion is powerful.
    B) that the lion is the king of the jungle.

6

Molly: A handbag is very useful to keep my cell phone and diary. That's why I like to carry it with me wherever I go.

Dolly: A handbag is useful to keep my lipstick and comb. That is why I like to carry it with me to all places.

What do Molly and Dolly agree about in their conversation?

  A) that they like to carry their cellphone wherever they go.
  B) that they like to carry a handbag with them wherever they go.

Verbal Reasoning   Answers-125
© Gift Of Logic, Inc * Copying prohibited

Name _____   Date _____

AGREE

7

Doctor: Since the baby has a fever, we need to give some medicine.

Nurse:. The baby has a high temperature. So, some medicine will definitely help.

What do the Doctor and the Nurse agree about?
    A) that the baby needs medicine.
    B) that the baby does not need medicine.

8

Pilot: This plane can seat 500 passengers. So, this airplane is very big.

Co-Pilot: This plane has very wide wings. That's why it is very big.

1) Do the Pilot and Co-Pilot agree that the plane is very big?
    A) Yes   B) No

2) Do the Pilot and Co-Pilot agree on why they think the plane is very big?
    A) Yes   B) No

Verbal Reasoning
© Gift Of Logic, Inc * Copying prohibited

Name _____   Date _____

DISAGREE

In this page, you will see someone expressing their opinion on something. You need to reply by disagreeing with that person.

1

Sarah: We should communicate clearly by talking very loudly.

Your reply:

2

Sanjay: Reading one story book every day is the best way to improve our vocabulary.

Your reply:

3

Jenny: Dora is boring. So, her birthday party will also be boring.

Your reply:

4

Justin: There is no use playing Chess.

Your reply:

Name _____   Date _____

DISAGREE

In this page, you will see someone speaking to you. You need to reply by disagreeing with whatever the person says.

5

Bonny: The temperature today is much colder than it was yesterday.

Your reply:

_____

6

Susan: Riding a bike to school is not good for your health.

Your reply:

_____

7

Farooq: Eating french fries is good for health because it is tasty.

Your reply.

_____

8

Kathy: We should not keep in touch with our old friends.

Your reply:

Name _____ Date _____

ANALYTICAL REASONING

Name _____ Date _____

1   LIST PROCESSING   position

A monkey, a squirrel, and a parrot sat from left to right in three positions of a branch numbered 1, 2 and 3 from left to right.

Write the names of the occupants in the list below.

Position	1	2	3
Occupant			

The parrot flew away and an eagle took its place instead. Now, modify the list to reflect this change.

Position	1	2	3
Occupant			

The squirrel jumped out of the branch and the monkey moved to take its place. Now, modify the list to reflect the change.

Position	1	2	3
Occupant			

Finally, the monkey and the eagle interchanged their positions. Modify the list to reflect this change.

Position	1	2	3
Occupant			

Analytical Reasoning   Answers-128
© Gift Of Logic, Inc * Copying prohibited

Name _____ Date _____

| **2** | **LIST PROCESSING**   add/remove |

Serena went to the grocery store to purchase the following list of things.

jam
peanut butter
bread
milk
orange juice

While she was in the grocery store, her mom called her and told her not to buy bread and jam, but buy green peas and popcorn. In the list below, write the items that Serena must buy now.

Name _____ Date _____

3 LIST PROCESSING   remove

The following list of flights were scheduled to arrive at an airport.

Flight Arrivals		
Flight#	From	Gate
121	Miami	22
133	Frankfurt	35
143	London	31
146	Mumbai	17

The flights from London and Miami ran into rough weather and had to be delayed. Fill in the tables below to show the new list of on-time flights
and delayed flights.

Flight Arrivals - On Time		
Flight#	From	Gate

Flight Arrivals - Delayed		
Flight#	From	Gate

Analytical Reasoning
© Gift Of Logic, Inc * Copying prohibited

| 4 | LIST PROCESSING | exclude |

Alex had a list of friends that he wanted to invite for his birthday party. That list is shown below.

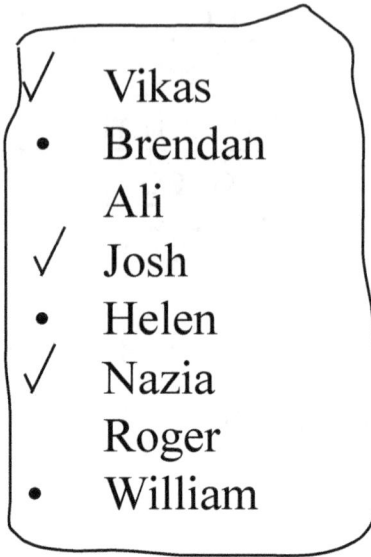

He called a few people yesterday and after inviting them, he marked his list with a check mark as shown above. His friends who are not in town have been marked with a dot (•). Make a new list showing only the friends that he must call today.

Name —————————————— Date ——————————————

5	LIST PROCESSING difference

Bruce Lee threw a birthday party. He asked his son Wang to prepare a list of all people at the party. He also asked his daughter Tang to do the same. Wang and Tang went around separately and wrote down the names of the people at the party. Their lists are shown below.

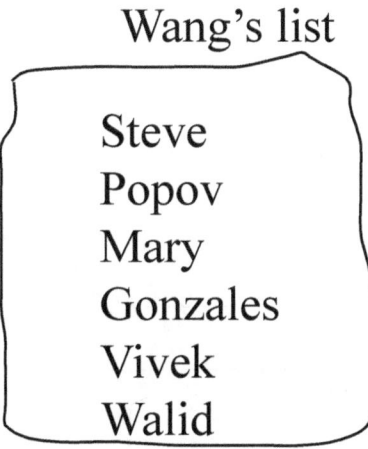
Wang's list

Steve
Popov
Mary
Gonzales
Vivek
Walid

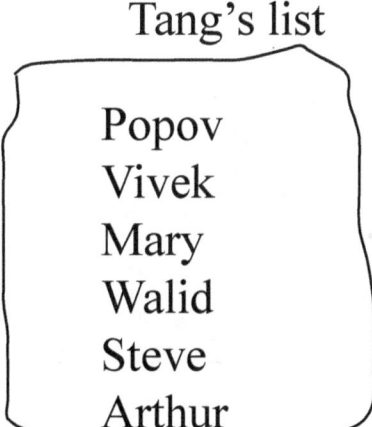
Tang's list

Popov
Vivek
Mary
Walid
Steve
Arthur

Based on the two lists shown above, answer the following questions.

1) Which of the following people did Wang not see at the party?
   A) Mary   B) Gonzales   C) Arthur

2) Which of the following people did Tang not see at the party?
   A) Arthur   B) Mary   C) Gonzales

3) If both Wang and Tang saw everyone who was at the party, how many people were at the party?

Analytical Reasoning   Answers-130
© Gift Of Logic, Inc * Copying prohibited

| 6 | LIST PROCESSING   alphabetic sort |

Sort the list in ascending order and write it in the empty list.

Popov
Vivek
Victor
Mary
Margie
Walid
Steve
Lu

Sort the list in descending order and write it in the empty list.

| Chris | Stacy | Mike | Steve | Miller |

Name _____   Date _____

7   LIST PROCESSING   sorting based on data

The following list shows the grades obtained by five students in a test. A grade of "A" means excellent and a grade of "E" means very poor.

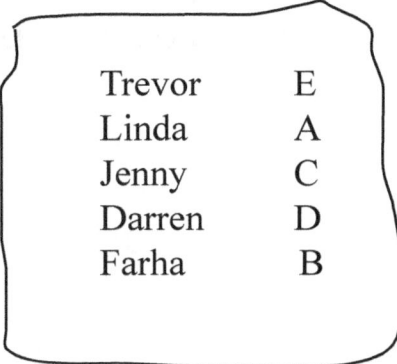

Trevor   E
Linda   A
Jenny   C
Darren   D
Farha   B

Sort the list by grade so that the student with the best grade is on the top. Sort the above list again so that the student with the worst grade is on the top. Complete the lists below.

best grade first

worst grade first

Analytical Reasoning
© Gift Of Logic, Inc * Copying prohibited

Name _____   Date _____

8 LIST PROCESSING   sorting based on data

numeric sort

The distance from the city of Dallas to nearby cities is shown below. Sort the list so that the city nearest to Dallas is on the top. Write the sorted list in the empty box shown and then answer the questions.

Houston	300
Waco	100
Richardson	15
Plano	18
Austin	200

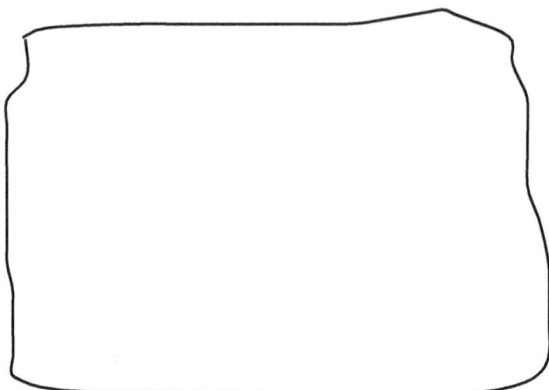

1) Which city is the nearest to Dallas?

2) Which city is the farthest from Dallas?

3) Which city is the second nearest to Dallas?

4) Which city is the third nearest to Dallas?

Analytical Reasoning
© Gift Of Logic, Inc * Copying prohibited

Name _____   Date _____

| 1 | SEQUENCING | rearrange |

Sam went to the school at 8 AM. He woke up at 6 AM. He returned home at 4 PM.

Rewrite the above sentences in sequence starting from the earliest event.

_____

A school hoists a flag of one color everyday. The schedule for flag hoisting is shown below.

> Thursday-Blue
> Wednesday-Green
> Tuesday- Red
> Monday- Orange
> Friday-Yellow

Rearrange the schedule in sequence starting from the first day of the week.

Name _____ Date _____

| **2** | SEQUENCING | reverse |

Write the months of the year from January in sequence.

First month to last month

Now, write the months of the year in reverse sequence starting from December. Do not refer to the list shown above.

Last month to first month

Analytical Reasoning
© Gift Of Logic, Inc * Copying prohibited

Name _____   Date _____

3	SEQUENCING

before/after

Alison went for a haircut in the morning after playing a game of soccer. After the haircut, she went home. While at home, she read a book and after that she played in the park. Later, she drew some pictures before going to bed.

Which of the following shows the correct sequence of activities of Alison from the beginning?
A) haircut, soccer, go home, read a book, play in the park, drew pictures, went to bed.
B) soccer, haircut, go home, read a book, play in the park, drew pictures, went to bed.

---

beginning/end
Ten people were standing in a line to get a free T-shirt.

Who was standing in front of the first person in the line?
Who was standing in front of the last person in the line?

---

reverse
Ten people were standing in a line. You must give a candy to each person consecutively starting with the tenth person.
1) After giving a candy to the fifth person, who will you give the candy to? A) Sixth person   B) Fourth person

2) After giving candy to the first person, who will you give a candy to?   A) Second person   B) Nobody

Analytical Reasoning   Answers-135   59
© Gift Of Logic, Inc * Copying prohibited

Name _____   Date _____

| **4** | SEQUENCING | alternate |

Number the rectangular boxes shown below sequentially beginning with 1. Also, name the balls between the boxes in alphabetic sequence starting with A.

☐ ○ ☐ ○ ☐ ○ ☐ ○ ☐ ○ ☐

Answer the following questions based on the sequence of objects that you see above.

1) Write the number sequence only:

2) Write the alphabetic sequence only:

3) Write the mixed sequence:

4) What are the first and last numbers in the sequence?

5) What are the first and last alphabets in the sequence?

6) What number is between alphabets C and D?

7) What alphabet is between numbers 3 and 4?

8) What alphabet is after number 6?

9) Which alphabet is before alphabet A?

10) Which number is after number 6?

Analytical Reasoning
© Gift Of Logic, Inc * Copying prohibited

Name _____ Date _____

1	GROUPING

Following is a group of animals. Split this group into two new groups - one group must have only domestic animals and the other group must have only wild animals.

Cow
Lion
Dog
Fox
Tiger
Horse

Domestic Animals   Wild Animals

1) How many animals are there in the Domestic Animals group?

2) How many animals are there in the Wild Animals group?

3) To which group will you add a cat? Write it in the correct list.

4) To which group will you add a cheetah? Write it in the correct list.

Analytical Reasoning   Answers-137
© Gift Of Logic, Inc * Copying prohibited

Name _____ Date _____

| 2 | GROUPING | grouping and sorting |

Grouping/Categorizing is done by gathering objects of similar type together.

Group the following list into vegetables and fruits.
  Grape, Carrot, Orange, Onion, Apple, Cauliflower

Vegetables	Fruits

After grouping the items, you can sort them either in ascending order or in descending order.

Sort the groups in ascending order and write them in the table below.

Vegetables	Fruits

Analytical Reasoning Answers-137
© Gift Of Logic, Inc * Copying prohibited

Name _____ Date _____

3 GROUPING sub-grouping

Tablets T1, T2, and T3 and syrups S1, S2, and S3 are used for different purposes as shown in the following table.

T1	Cold
T2	Headache
T3	Cough
S1	Cough
S2	Headache
S3	Cold

0) Which medicine used for treating headaches is a syrup?

1) Write the group of medicines used for treating cold.

2) Write the group of medicines used for treating cough.

3) Write the group of medicines used for treating cold and cough.

4) Write the group of medicines used for treating headache and cold.

Analytical Reasoning   Answers-138
© Gift Of Logic, Inc * Copying prohibited

Name —————————————— Date ——————————————

VENN DIAGRAM BASICS

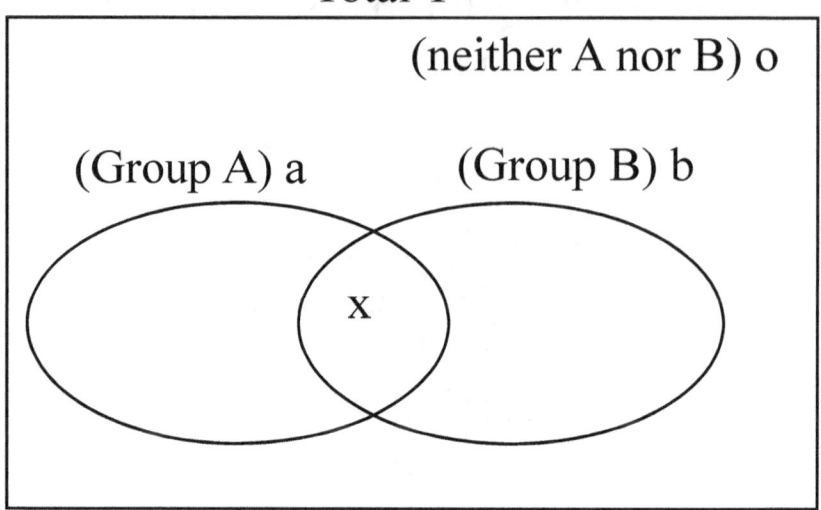

Venn Diagrams are used to represent the relationships between two or more groups. In this section, we will consider two groups only. When you classify a total of T items into two groups A and B, there can be items that belong to group A only, items that belong to group B only, items that belong to both group A and group B, and items that belong to neither group A nor group B.

In the Venn diagram, the ellipses that represent group A and group B are shown. There are $\underline{a}$ items in group A, and $\underline{b}$ items in Group B. The area common to groups A and B represents the items that belong to both groups. The number of items in this common area is represented by $\underline{x}$. There are $\underline{o}$ items that belong to neither group A nor group B. Now, we can answer some interesting questions about these groups.

1) How many items are in group A only?

As you can see, there are $\underline{a}$ items in group A. Of these $\underline{a}$ items, $\underline{x}$ items also belong to group B. So, only a-x items belong to group A only.

Analytical Reasoning
© Gift Of Logic, Inc * Copying prohibited

Name ────────────── Date ──────────────

VENN DIAGRAM BASICS

2) How many items are in group B only?

As you can see, there are <u>b</u> items in group B. Of these <u>b</u> items, <u>x</u> items also belong to group A. So, a total of b-x items belong to group B only.

3) How many items belong to both group A and group B.

There are <u>x</u> items that belong to both group A and group B. Now, the Venn diagram can be completed as shown below.

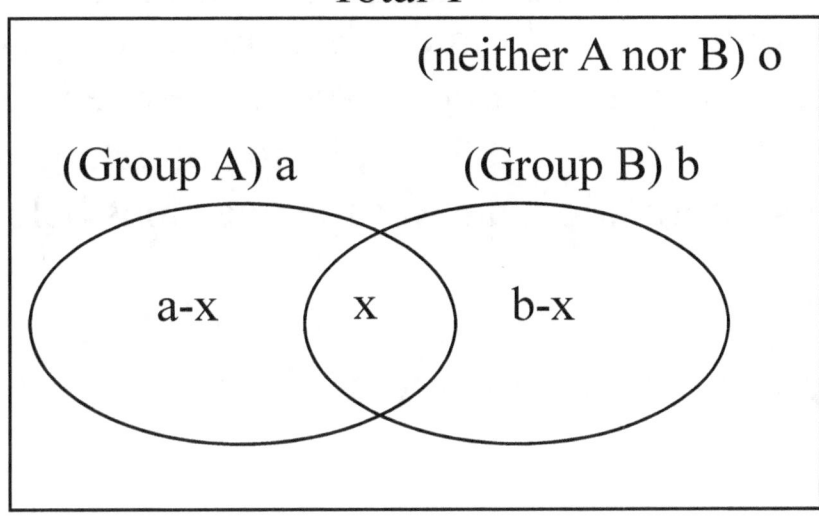

4) How many items belong to group A or group B?
   a-x items belong to group A only.
   x items belong to both group A and group B.
   b-x items belong to group B only.
So, a total of a-x + x + b-x = a+b-x items belong to group A or B.

5) How many total items are there?
   Add up all the items that are being considered.
   group A items + group A and B items + group B items + neither group A nor group B items = a-x+x+b-x+o = a+b-x+o = T

Analytical Reasoning
© Gift Of Logic, Inc * Copying prohibited

1 VENN DIAGRAM

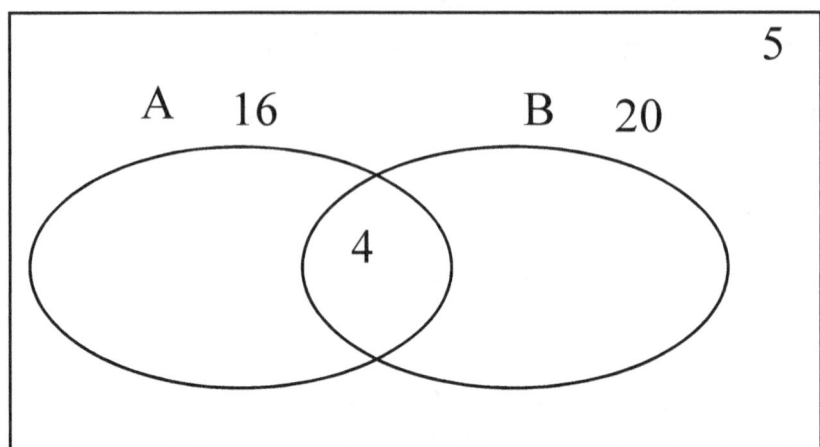

All the people in a room were split into two groups, group A and group B. There are 16 people in group A, and 20 people in group B. 4 people belong to both groups and 5 people do not belong to either group. Based on this information, answer the questions below.

1) How many people are there in the room?

2) How many people are there in either group?

3) How many people belong to group A only?

4) How many people belong to group B only?

Analytical Reasoning
© Gift Of Logic, Inc * Copying prohibited

Name _____   Date _____

2 VENN DIAGRAM

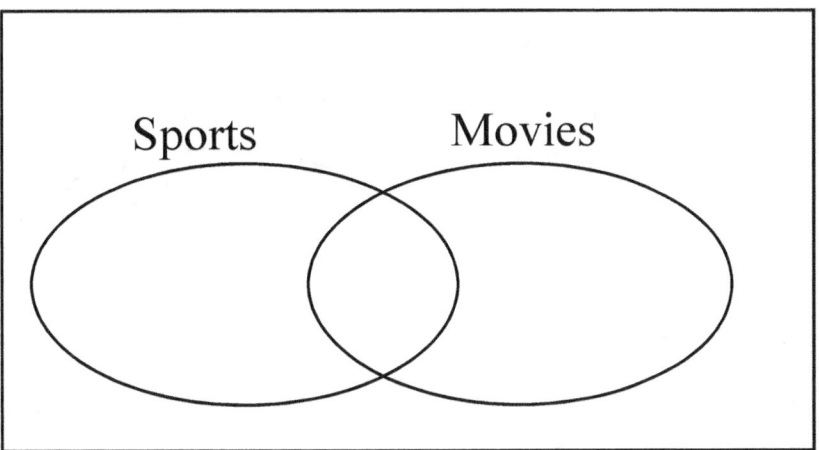

25 students were surveyed and asked if they liked sports or movies. 15 said that they liked sports, 8 said that they liked movies and 5 said that they liked neither. Complete the Venn diagram and answer the following questions.

1) How many students surveyed liked sports as well as movies?

2) How many students liked sports only?

3) Ho many students liked movies only?

Analytical Reasoning
© Gift Of Logic, Inc * Copying prohibited

| 3 | VENN DIAGRAM |

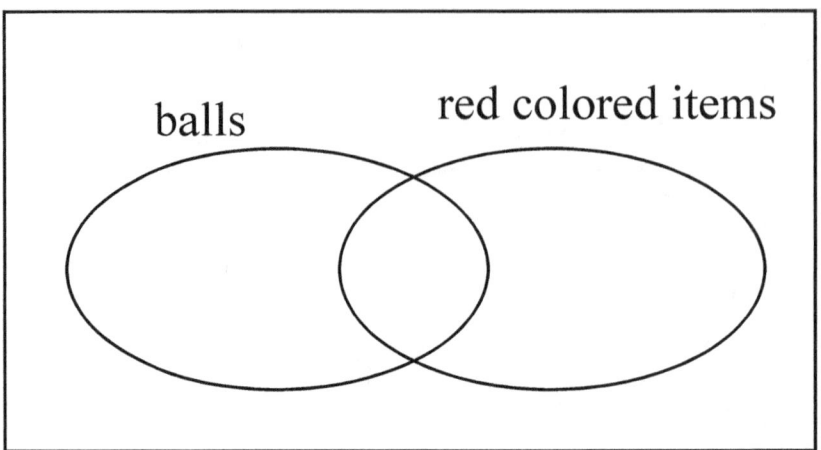

There are fifty items in a box. 25 of the items are balls. 30 items are red colored items. 10 items are not either balls or red colored. Complete the Venn diagram and answer the questions below.

1) How many items in the box are either balls or red colored?

2) How many items are red colored and balls?

3) How many items are not balls?

4) How many items are not red items?

5) How many items are neither red nor balls?

| 1 | GRAPH LOGIC |

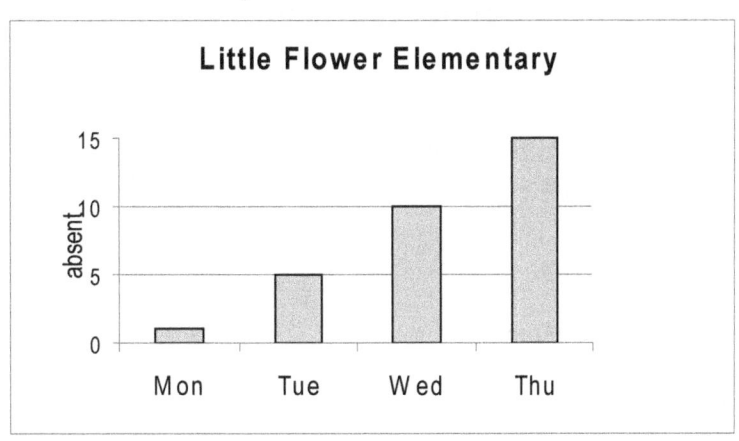

The graph above shows the number of students absent for school during one particular week. Read the graph carefully and answer the questions below.

1) More students were absent on Wednesday than on Tuesday.
   A) True   B) False

2) More students attended school on Thursday than on Tuesday.
   A) True   B) False

3) From Monday to Thursday, the number of students attending school increased.
   A) True   B) False

| **2** | **GRAPH LOGIC** |

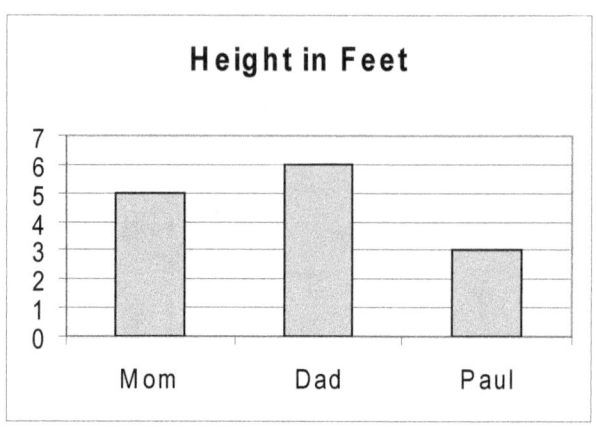

The graph above shows the height in feet of the members of Paul's family. Read the graph carefully and answer the questions below.

1) Who is the tallest in Paul's family?
   A) Mom   B) Dad   C) Paul

2) Who is the shortest in Paul's family?
   A) Mom   B) Dad   C) Paul

3) If Paul was three feet taller than what he is now, he will be taller than his dad.
   A) True   B) False

4) If Paul was three feet taller than what he is now, he will be taller than his mom.
   A) True   B) False

| **3** | **GRAPH LOGIC** |

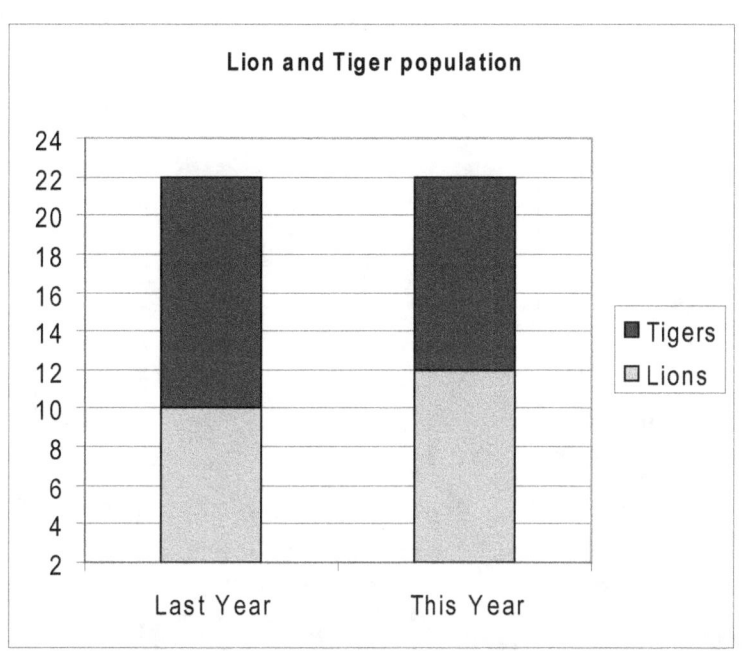

The graph above shows the Lion and Tiger population in a wild life reserve. Read the graph carefully and answer the following questions.

1) There were more Lions last year than there were Tigers.
   A) True   B) False

2) There are more Tigers this year than Lions.
   A) True   B) False

3) There were the same number of Lions last year as there are Tigers this year.
   A) True   B) False

Name _____ Date _____

4 GRAPH LOGIC

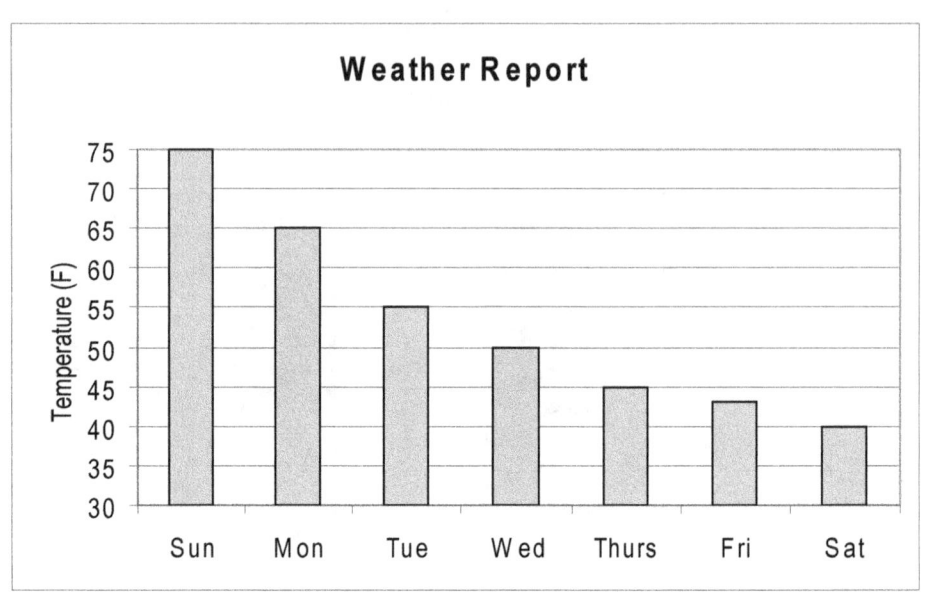

The weather for one full week for a city is shown in the graph above. Read the graph carefully and answer the questions.

1) The temperature increased from 40 degrees to 75 degrees from Sunday to Thursday.
    A) True   B) False

2) Temperatures remained constant for three days in a row.
    A) True   B) False

3) Temperatures fell more from Sunday to Tuesday than from Tuesday to Thursday.   A) True   B) False

Name _____ Date _____

| 5 | GRAPH LOGIC |

Three parks, A, B, and C had the following number of visitors during different seasons as shown in the following chart.

	Park A	Park B	Park C
Spring	20	40	20
Summer	30	30	30
Winter	40	30	40

Read the chart carefully and answer the questions below.

1) Park-A had the same number of visitors as Park-B and Park-C in spring and summer.

      A) True   B) False

2) Park-A had the same number of visitors as Park-C in summer and winter.

      A) True   B) False

Analytical Reasoning   Answers-145

© Gift Of Logic, Inc * Copying prohibited

GRAPH LOGIC

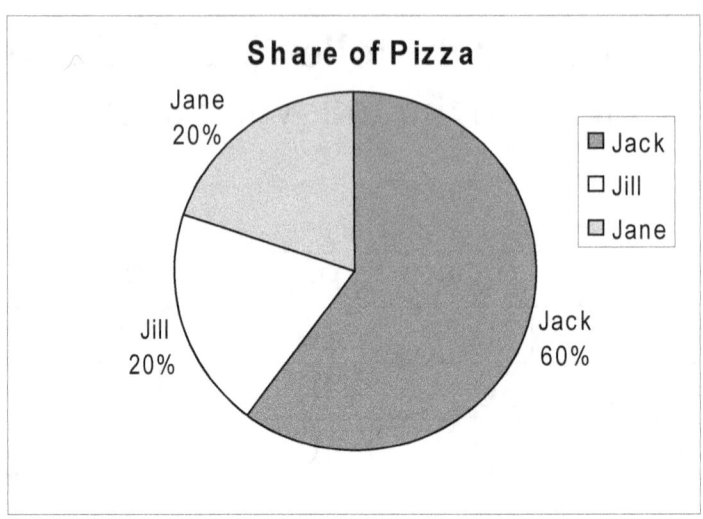

Jack, Jill, and Jane ate pizza during a birthday party. Their individual shares are shown in the pie chart above. Read the graph carefully and answer the questions below.

1) Jack and Jill ate the same amount of pizza.
   A) True   B) False

2) Jane and Jill ate the same amount of pizza.
   A) True   B) False

3) Jane and Jill ate the same number of pizza slices.
   A) True   B) False   C) Uncertain

7	GRAPH LOGIC

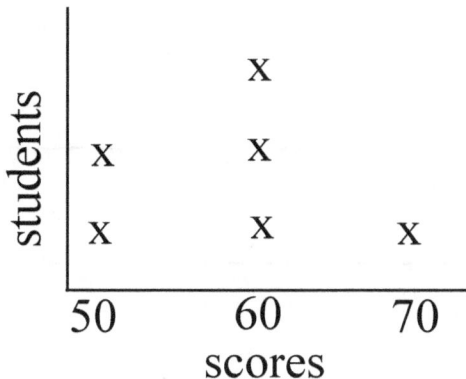

The graph above shows the scores obtained by students in a math test. Each student is represented by the symbol x. Read the graph carefully and answer the questions below.

1) Half the number of students scored more than 60 points.
   A) True   B) False

2) One third of the students have the lowest score.
   A) True   B) False

Name _____ Date _____

NUMBER LOGIC

Figure out the logic in the number sequence and write the missing number.

1	1 3 5 ?
2	2 5 8 ?
3	11 22 33 ?
4	10 20 ? 40
5	? 7 6 5
6	10 100 1000 ?
7	10000 1000 ? 10
8	0.5 1.5 2.5 ?
9	9.5 7 ?
10	0.25 0.75 2.25 ?

Analytical Reasoning   Answers-147
© Gift Of Logic, Inc * Copying prohibited

Name _____   Date _____

NUMBER LOGIC

Figure out the logic in the numbers and find the missing number.

11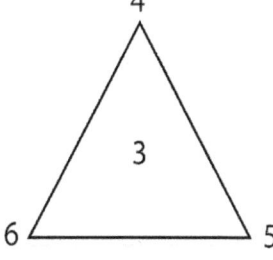

12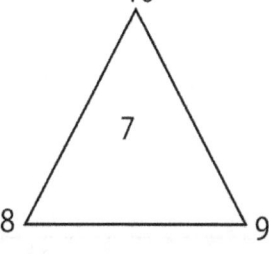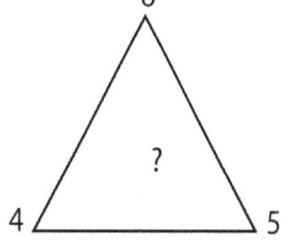

13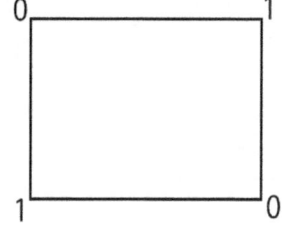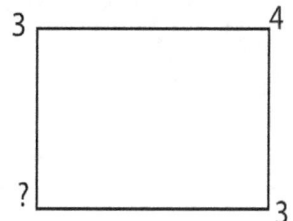

14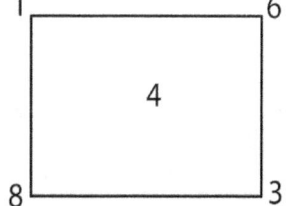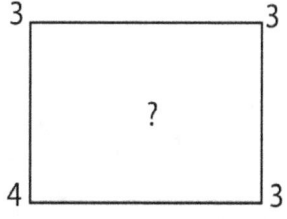

15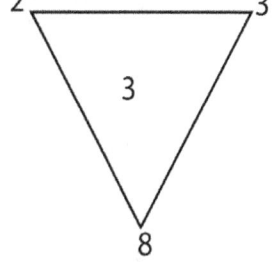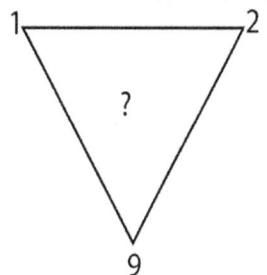

Analytical Reasoning
© Gift Of Logic, Inc * Copying prohibited

Name _____ Date _____

NUMBER LOGIC

Figure out the logic in the numbers and find the missing number.

16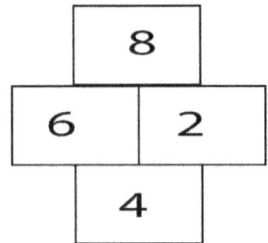

17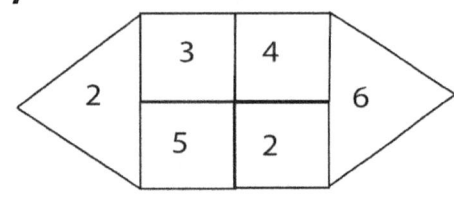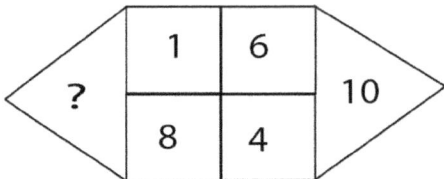

18

19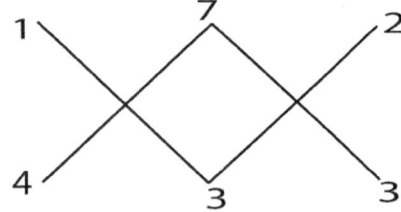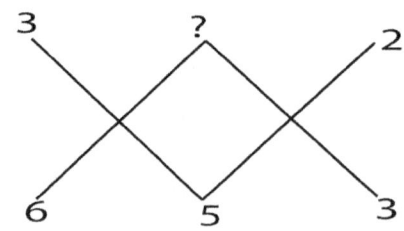

20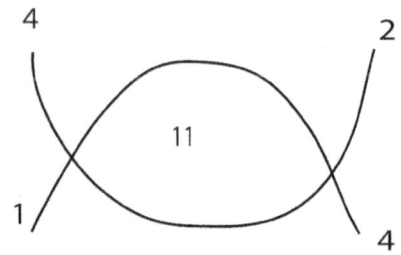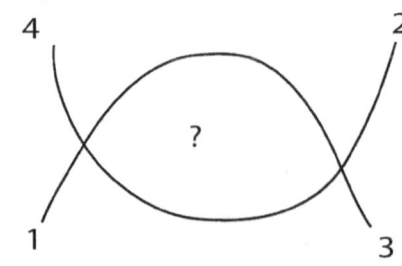

Analytical Reasoning

LETTER LOGIC

Figure out the logic in the sequence and find the missing letter or number.

1	A C E ?
2	K ? Q T
3	Z Y X ?
4	Z X V ?
5	A 1 B 2 C ?
6	AB CD ? GH
7	1 A 2 B ? C
8	Z1 Y2 ? W4

Analytical Reasoning   Answers-150

Name ——————————————— Date ———————————

LETTER LOGIC

Figure out the logic in the sequence and find the missing letter or number.

9
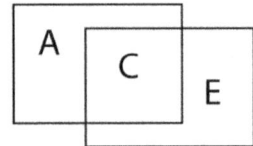 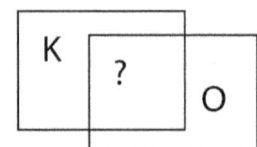

10
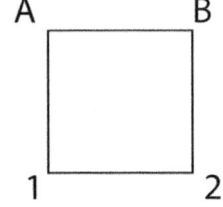 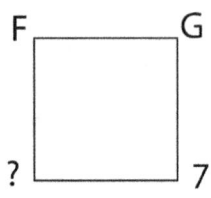

11
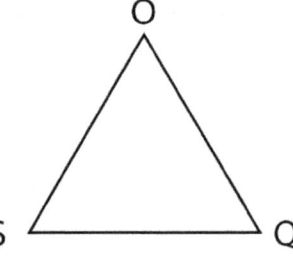 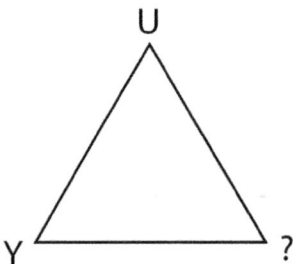

12
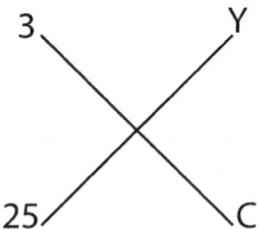 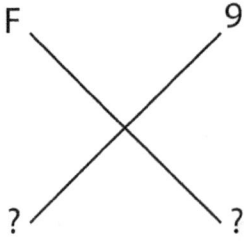

Analytical Reasoning Answers-151
© Gift Of Logic, Inc * Copying prohibited

LETTER LOGIC

Figure out the logic in the sequence and find the missing letters.

13

```
    A           K           U
D   B    N     L    X    ?
    C           M           W
```

14

AB   ZY   CD   ?

15

AC   BD   CE   ?

16

BA   DC   FE   ?

Analytical Reasoning

Name _____ Date _____

1. SUDOKU

Solve the following Sudoku. A correctly solved Sudoku has numbers 1-9 appearing only once in each row, each column and each 3x3 grid. You gain valuable positioning skills by solving these sudokus.

1	5		9	7		4	2	6
7		4	2	3	6		1	5
	6	9	5		1	7	3	8
5	4	1	7	6	3	2	8	
8		6	4	2	9		5	7
9	2	7		1	5	3		4
3	7	5	1	8	4	6	9	2
	9	2	3		7	8	4	
4	1	8	6	9	2	5	7	3

Analytical Reasoning
© Gift Of Logic, Inc * Copying prohibited

Name _____ Date _____

2 SUDOKU

Solve the following Sudoku. A correctly solved Sudoku has numbers 1-9 appearing only once in each row, each column and each 3x3 grid. You gain valuable positioning skills by solving these sudokus.

4	3	1	6		9	8	5	
2		5	8		3	9		7
8	9		4	5	2	3	1	6
7	4	2		9	1	6	3	8
	8	6	2	4		5	9	1
1	5	9	3	6	8			4
9	2	8	1	3		7	6	
6	7			8	5	1	2	3
5	1	3	7	2	6		8	9

Analytical Reasoning
© Gift Of Logic, Inc * Copying prohibited

Name _____   **Date** _____

3
SUDOKU

Solve the following Sudoku. A correctly solved Sudoku has numbers 1-9 appearing only once in each row, each column and each 3x3 grid. You gain valuable positioning skills by solving these sudokus.

3	7	4	8		9	2	5	
	9	1	3		5	4	8	6
5	8	6		4	2		9	7
	1	3	4	8	7		6	2
6	4		5		1	7	3	
7		2	9	3	6	8	1	
1	3	7	2	9	8	6		5
8	2		6	5		1		3
4	6	5	7	1	3	9	2	8

Analytical Reasoning
© Gift Of Logic, Inc * Copying prohibited

Name _____ Date _____

4	SUDOKU

Solve the following Sudoku. A correctly solved Sudoku has numbers 1-9 appearing only once in each row, each column and each 3x3 grid. You gain valuable positioning skills by solving these sudokus.

	7	1	5	6		4	3	8
8		4	2	7	1	9	5	6
5	6	9	3	8		2		1
9	5	3	8		7	6	1	2
7	8		1	5	6			9
	1	6	9	2	3	5	8	7
3		7	6		8	1	2	5
1	9	5		3	2	8	6	
6	2		4	1	5	7	9	3

Analytical Reasoning
© Gift Of Logic, Inc * Copying prohibited

SUDOKU

5

Solve the following Sudoku. A correctly solved Sudoku has numbers 1-9 appearing only once in each row, each column and each 3x3 grid. You gain valuable positioning skills by solving these sudokus.

4	1	7	5		2	8		9
9		6	4	3	8	7	1	
	3	5	7	9	1			6
5	6	8	3		9	2	7	4
2		1		7	6		5	3
7		3	2		4	1	6	
6		4	1		5	3		2
	8	9	6		7	5		1
1		2		4	3		8	7

Analytical Reasoning
© Gift Of Logic, Inc * Copying prohibited

Name ——————————————— Date ———————————————

PICTORIAL REASONING

Pictorial Reasoning
© Gift Of Logic, Inc * Copying prohibited

Name ——————————   Date ——————————

PICTURE SEQUENCE

Figure out the logic in the picture sequence, and draw the next picture in the sequence.

1   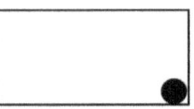 ?

2 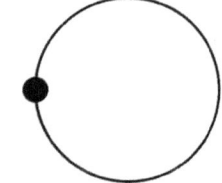  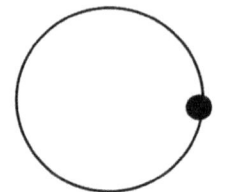 ?

3   ?

4  ?

5 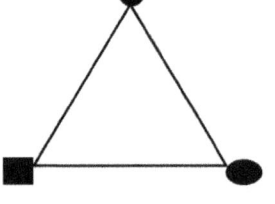 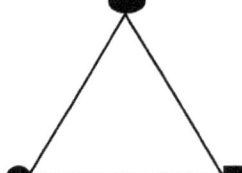 ?

6  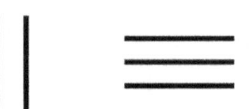 ?

7 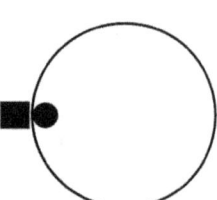 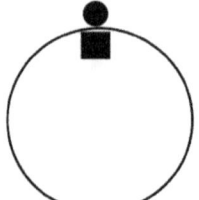 ?

Pictorial Reasoning   Answers-158   88
© Gift Of Logic, Inc * Copying prohibited

Name ____________________ Date____________________

PICTURE SEQUENCE

Figure out the logic in the picture sequence, and draw the next picture in the sequence.

8

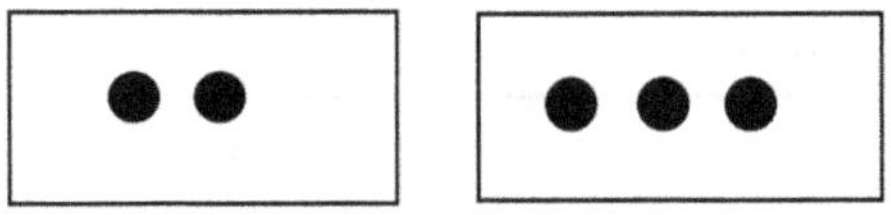

?

9

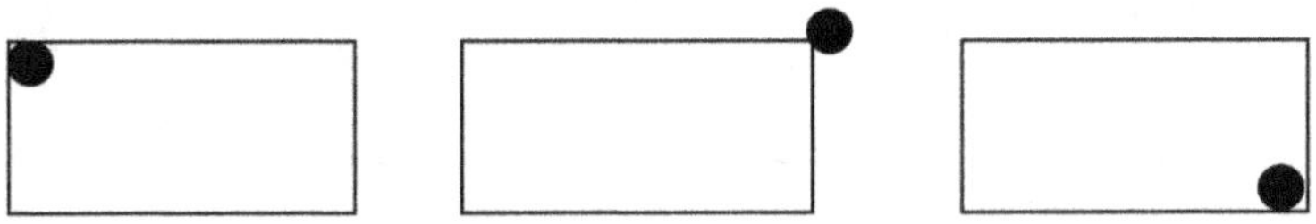

?

10

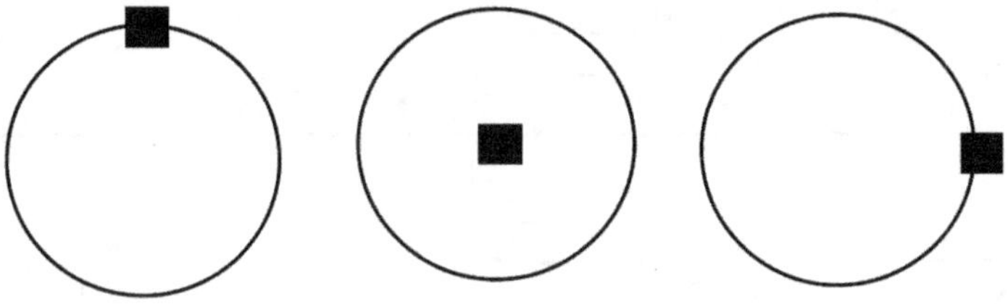

?

11

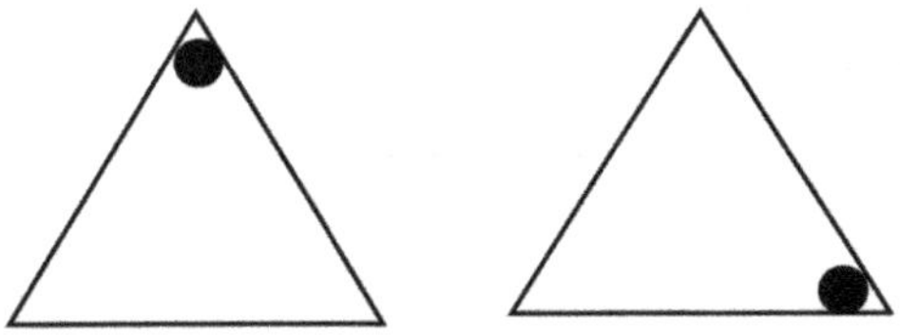

?

12

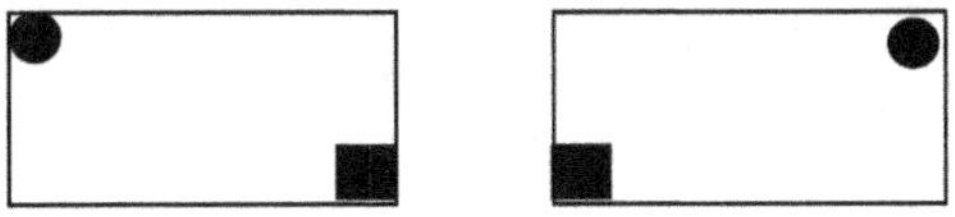

?

Name _____   Date _____

PICTURE SEQUENCE

Figure out the logic in the picture sequence, and draw the next picture in the sequence.

13 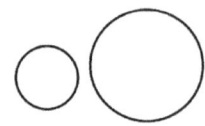      ?

14 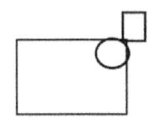   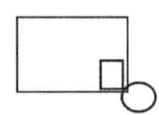   ?

15 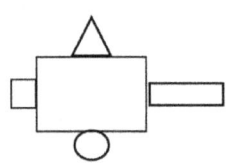   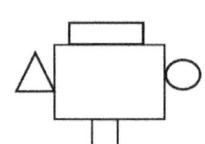   ?

16 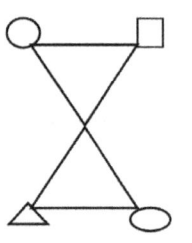   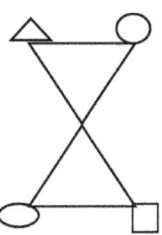   ?

17 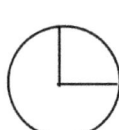   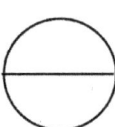   ?

18 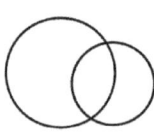   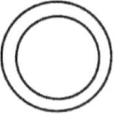   ?

Pictorial Reasoning   Answers-160
© Gift Of Logic, Inc * Copying prohibited

Name ———————————— Date ————————————

PICTURE ANALOGY

Figure out the logic in the picture analogy, and circle the correct picture that will complete the analogy.

1

A   B   C

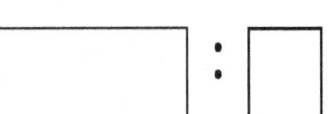 AS  :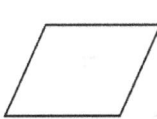

2

A   B   C

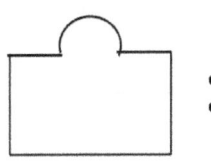 AS  :

3

A   B   C

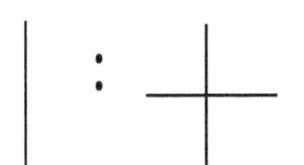 AS  :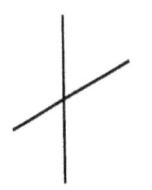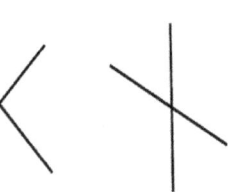

Pictorial Reasoning   Answers-161
© Gift Of Logic, Inc * Copying prohibited

Name _____ Date _____

PICTURE ANALOGY

Figure out the logic in the picture analogy, and circle the correct picture that will complete the analogy.

4

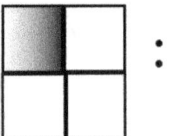 : 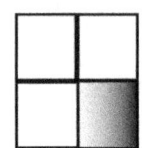 AS 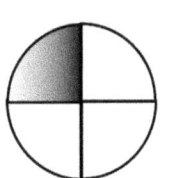 :   A   B   C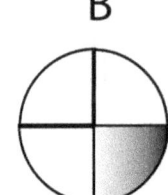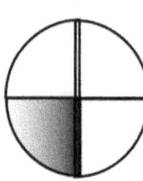

5

 AS   A   B   C 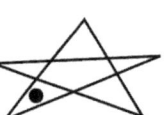 :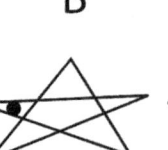

6

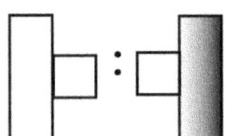 AS   A   B   C 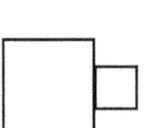 :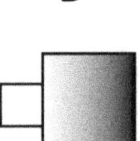

Pictorial Reasoning

Name _____   Date _____

PICTURE ANALOGY

Figure out the logic in the picture analogy, and draw the picture that will complete the analogy.

7 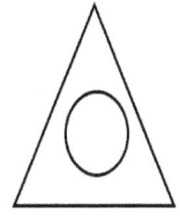 :  AS 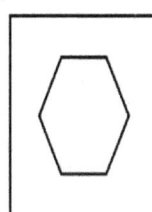 : ?

8  : AS 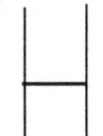 : ?

9  : AS  : ?

10 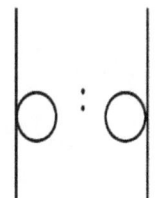 AS 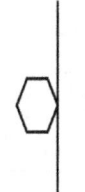 : ?

Pictorial Reasoning
© Gift Of Logic, Inc * Copying prohibited

Name _____ Date _____

ODD PICTURE

In each question below, find the odd picture and circle the answer.

1 A B C

 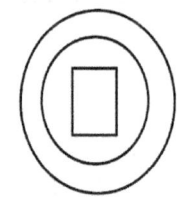 

2 A B C

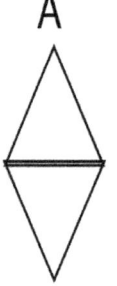  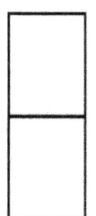

3 A B C

4 A B C

```
V V V V    U U U U    T T T T
 V V        U U U      T T
 V V         U          T T
  V          U           T
```

5 A B C

 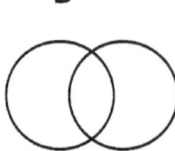 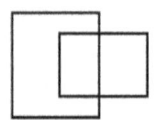

Pictorial Reasoning Answers-164 94
© Gift Of Logic, Inc * Copying prohibited

Name _____   Date _____

ODD PICTURE

In each question below, find the odd picture and circle the answer.

6 A B C

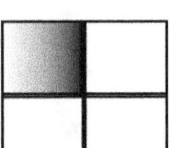 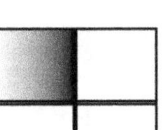 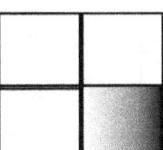

7 A B C

8 A B C

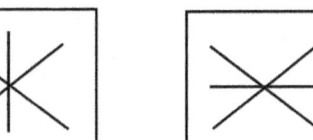 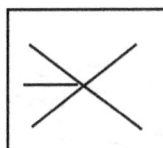

9 A B C

 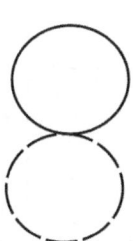

10 A B C D

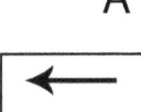 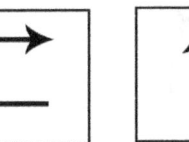 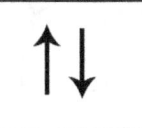 

Pictorial Reasoning   Answers-165

© Gift Of Logic, Inc * Copying prohibited

Name —————————————— Date ——————————————

PICTURE DIFFERENCE

Mark the differences in the set of pictures shown, with arrows.

1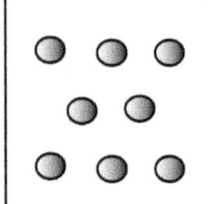

2

3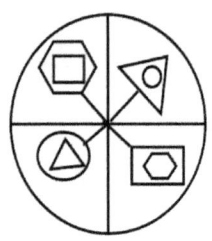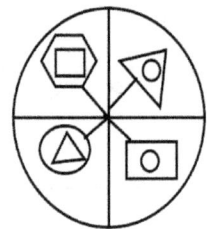

4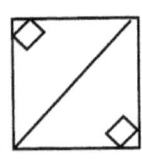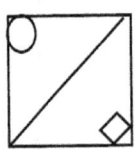

5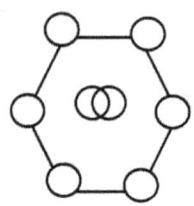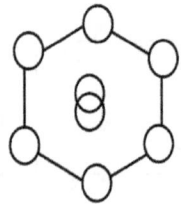

Pictorial Reasoning Answers-166
© Gift Of Logic, Inc * Copying prohibited

Name ——————————— Date ———————————

PICTURE DIFFERENCE

Mark the differences in the set of pictures shown, with arrows.

6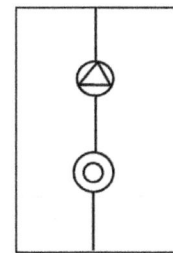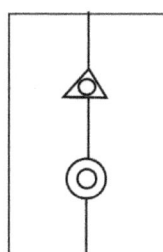

7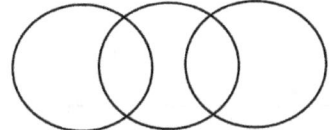

8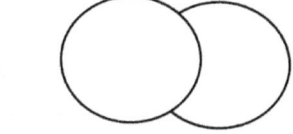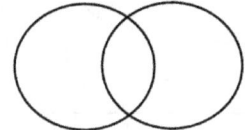

9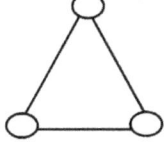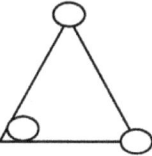

10

Pictorial Reasoning
© Gift Of Logic, Inc * Copying prohibited

Name _____ Date _____

PICTURE DIFFERENCE

Mark the differences in the set of pictures shown, with arrows.

11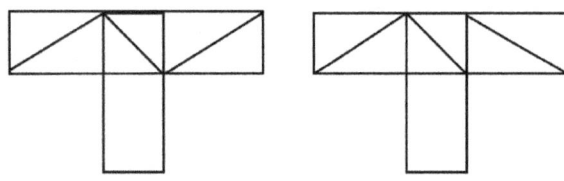

12

13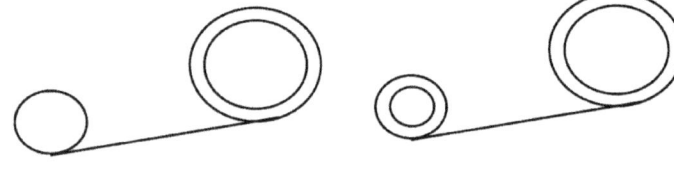

14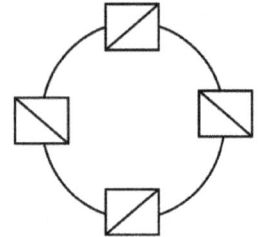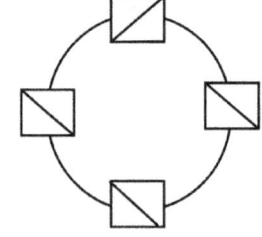

15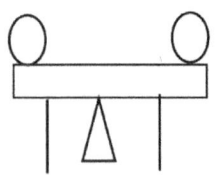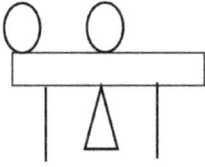

Pictorial Reasoning

Name _____ Date _____

PATTERN MATCHING

Find the pattern in the picture-set on the left, and identify the picture on the right that will fit in the space marked with ? to complete the pattern.

1 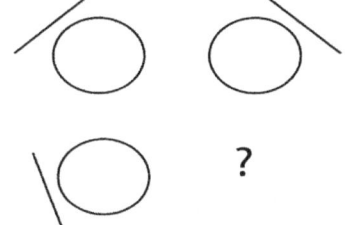   A   B   C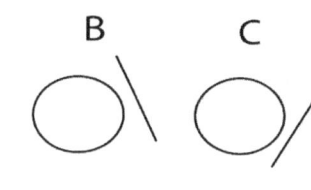

2 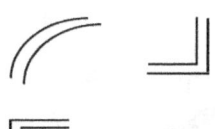   A   B   C

3    A   B   C

4 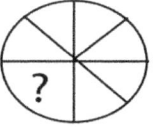   A   B   C

5 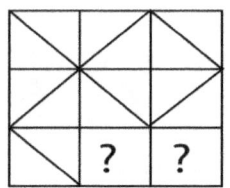   A   B   C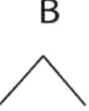

Pictorial Reasoning
© Gift Of Logic, Inc * Copying prohibited

Answers-169

99

ANSWERS

Logical Negation using NOT

Negate the following statements.

1 Statement: World War II happened in 1990.
Negation: World War II did not happen in 1990.
Negation is A) True

2 Statement: Airplanes travel in the ocean.
Negation: Airplanes do not travel in the ocean.
Negation is A) True

3 Statement: Mercury is the planet closest to the Sun.
Negation: Mercury is not the planet closest to the Sun.
Negation is B) False

4 Statement: Wild animals are not found in the zoo.
Negation: Wild animals are found in the zoo.
Negation is A) True

5 Statement: Computers cannot play chess.
Negation: Computers can play chess.
Negation is A) True

6 Statement: All doctors are women.
Negation: All doctors are not women.
Negation is A) True

Answers
© Gift Of Logic, Inc * Copying prohibited

Symbolic Representation of NOT (~)

A negation can be represented in logical form using the NOT symbol ~.
Doing so helps to represent the facts in a concise format that can be used for making inferences.

In the statements below, negate the underlined portion using the ~ symbol.
Example:

1 Statement: World War II <u>happened in 1990.</u>
Symbolic Negation: ~happen in 1990

Airplanes <u>travel in the ocean.</u>
Symbolic Negation: ~travel in the ocean

---

2 Statement: Mercury is the planet <u>closest</u> to the Sun.
Symbolic Negation: ~closest

---

3 Statement: Wild animals are <u>not found in the zoo.</u>
Symbolic Negation: ~not found in the zoo
Note: ~not found in the zoo is the same as "found in the zoo".

---

4 Statement: Computers <u>cannot play chess.</u>
Symbolic Negation: ~cannot play chess
Note: ~cannot play chess is the same as "can play chess"

---

5 Statement: <u>All doctors are women.</u>
Symbolic Negation: ~all doctors are women

Logical Conjunction using AND

In the following exercises, find the truth of the conjunction.

1 Conjunction: Some apples are green and some apples are blue. Conjunction is B) False <u>Reasoning:</u> The first conjunct, "some apples are green" is true. The second conjunct "some apples are blue" is false. If one conjunct is false, the entire conjunction is false. True and False is False.

2 Conjunction: Human beings can jump and fly. Conjunction is B) False <u>Reasoning:</u> Human beings can jump, but cannot fly. True and False is False.

3 Conjunction: Cotton can catch fire and cotton is not soft. Conjunction is B) False <u>Reasoning:</u> Cotton is not soft is a false statement. True and False is False.

4 Conjunction: Some buildings are tall and some buildings are short.   Conjunction is A) True <u>Reasoning:</u> It is true that some buildings are tall and some buildings are short. Since both conjuncts are true, the entire conjunction is true. True and True is True.

5 Conjunction: A pen is a musical instrument and so is a guitar. Conjunction is B) False <u>Reasoning:</u> False and True is False.

6 Conjunction: A car has two wheels and a bike has no wheels. Conjunction is B) False <u>Reasoning:</u> False and False is False.

Answers
© Gift Of Logic, Inc * Copying prohibited

Symbolic Representation of Conjunction (&)

Represent the following conjunctions using &.

1 Conjunction: A car has four wheels and a bike has no wheels.
Symbolic: car has four wheels & bike has no wheels

2 Conjunction: Human beings can jump and fly.
Symbolic: jump & fly

3 Conjunction: Cotton can catch fire and cotton is not soft.
Symbolic: catch fire & not soft

4 Conjunction: She ranked first in singing and third in swimming.
Symbolic: first in singing & third in swimming

5 Conjunction: A pen and a guitar are musical instruments.
Symbolic: pen & guitar

Answers
© Gift Of Logic, Inc * Copying prohibited

Logical Disjunction   Inclusive OR - Exclusive OR

1 A tiger can live in a forest or a zoo.

The "OR" in the statement is an A) inclusive OR
Reasoning: A tiger can live in forest or a zoo or both.

Tigers can live in
 C) a forest, a zoo or both

---

2 John can be in team A or team B or both.

The "OR" in the statement is an A) inclusive OR
Reasoning: The statement says that John can belong to either team or he can belong to both teams. So, it is an inclusive OR.

John can belong to which of the following teams? C) A and B
Reasoning: John can belong to both teams.

---

3 Lisa can be in team A or team B, but not both.

The "OR" in the statement is an B) exclusive OR
Reasoning: The statement clearly excludes the possibility of Lisa belonging to both teams with "but not both".

Lisa can belong to which of the following teams?   A) A   B) B
Reasoning: Lisa can belong to either team A or team B, but not both.

Answers

Logical Disjunction   Inclusive OR - Exclusive OR

4 Mary can sit in the first chair or the fourth chair now.

The "OR" in the statement is an B) exclusive OR
Reasoning: Since a person can sit in only one chair at a time, Mary can sit in the first chair or the fourth chair, but not both. So, the "or" in the statement is an exclusive OR.

Mary can sit in which of the following chairs?
Answer: B) first or fourth

Reasoning: Mary can be sit in either of the chairs only, but not in both chairs.

---

5   Tom can eat an apple or an orange, but not both.

The "OR" in the statement is an B) exclusive OR
Reasoning: Since the statement clearly states "..but not both", it implies that only one of the disjuncts can be true.

At most, how many fruits can Tom eat?   B) 1

Reasoning: Tom can eat an apple or an orange, but he cannot eat both. So, he can eat at most one fruit.

> # Symbolic Representation of Logical Disjunction - OR (‖) (╫)

Represent the following statements by using the symbolic notation for OR.
Use the symbols ‖ for inclusive OR and ╫ for exclusive OR.

---

1 Disjunction: John can be in team A or team B or both.
Symbolic: team A ‖ team B

---

2 Disjunction: Lisa can be in team A or team B, but not both.
Symbolic: team A ╫ team B

---

3 Disjunction: Mary can sit in the first chair or the fourth chair.
Symbolic: first chair ╫ fourth chair

---

4 Disjunction: Tom can eat an apple or an orange.
Symbolic: apple ‖ orange

---

5 Disjunction: The tea is served either hot or cold.
Symbolic: hot ╫ cold

Answers

© Gift Of Logic, Inc * Copying prohibited

Symbolic Representation of if P ... then Q (→)

A conditional statement "If P then Q" can be symbolically represented as

P → Q

Conditional: If Martha dances, then Mark will sing.
Symbolic Representation: Martha dances → Mark will sing.

Represent the following conditional statements in symbolic form. Identify the antecedent and consequent as well.

---

1 Conditional: If Martha cries, then Mark will smile.

Symbolic: Martha cries → Mark will smile
Antecedent: Martha cries   Consequent: Mark will smile

---

2 Conditional: If the baby is hungry, it will cry.

Symbolic: baby is hungry → it will cry

Antecedent: baby is hungry   Consequent: it will cry

---

3 Conditional: Mark will sing if Martha dances.

Symbolic: Martha dances → Mark will sing

Antecedent: Martha dances   Consequent: Mark will sing
Note the position of "if" in the conditional and note the position of the antecedent and the consequent.

Answers

© Gift Of Logic, Inc * Copying prohibited

CONDITIONAL STATEMENTS   If P then Q using "Not"

1 Write a conditional statement with "not" in its antecedent. Represent the statement in symbolic form.

If you are not responsible, you will suffer.
Symbolic: ~responsible → suffer

---

2 Write a conditional statement with "not" in its consequent. Represent the statement in symbolic form.

If you water the plants, the plants will not die.
Symbolic: water the plants → ~die

---

3 Write a conditional statement with "not" in both its antecedent and consequent.

If you do not have enough money, you will not be able to buy the toy.
Symbolic: ~enough money → ~buy the toy

Answers

CONDITIONAL STATEMENTS - Unless

Convert the following conditional statements that use "Unless" into conditional statements and write their symbolic form.

---

1 Unless: Unless it stops raining, the shuttle will not take off.

If..then: If it does not stop raining, the shuttle will not take off.
Symbolic: ~stop raining → ~take off

---

2 Unless: Unless everyone is silent, the show will not begin.

If..then: If everyone is not silent, the show will not begin.
Symbolic: ~everyone silent → ~begin

3

Unless: Unless you exercise, you will not be healthy.

If..then: If you do not exercise, you will not be healthy.
Symbolic: ~exercise → ~healthy

---

4 Unless: You must not leave unless I come back from shopping.

If.. then: You must not leave if I do not come back from shopping.
Symbolic: ~come back → ~leave

Answers

© Gift Of Logic, Inc * Copying prohibited

CONDITIONAL STATEMENTS - Except

Convert the following conditional statements that use "Except" into conditional statements and write their symbolic form.

1

Except: Except on Wednesdays, you must wear a tie every day.
If.. then: If it is not a Wednesday, you must wear a tie.
Symbolic: ~Wednesday → wear tie

2

Except: Except during winter, you must walk to school.
If..then: If it is not winter, you must walk to school.
Symbolic: ~winter → walk to school

3

Except: Everyone must practice at 6 PM except George.
If..then: If you are not George, you must practice at 6 PM.
Symbolic: ~George → must practice at 6

4

Except: The park must be closed for public except when it is summer.
If..then: If it is not summer, then the park is closed for public.
Symbolic: ~summer → closed for public

5

Except: Except on Mondays, the recreation center will be open everyday.
If..then: If it is not a Monday, the recreation center is open.
Symbolic: ~Monday → open

Answers

INFERENCING-must be true

1 Brandi's mom said that.. Answer: A) She saw a clown.
<u>Reasoning:</u> First, note the condition in the statement and write it in symbolic form. go to the circus → can see a clown
Since she went to the circus, it must be true that she saw a clown. Remember that if the antecedent happens, then the consequent will happen.

---

2 Andy's grandma: "If you try hard,...
Answer: A) Andy was able to climb Mount Everest.
<u>Reasoning:</u> The conditional statement is:
   try hard → will be able to climb Mount Everest
Since Andy tried hard, it must be true that he was able to climb Mount Everest.

---

3 If the bell rings, the students.. Answer: B) The students went home. <u>Reasoning:</u> The conditional statement is:
   bell rings → students will go home
Since the bell rang, it must be true that the students went home. Based on this condition alone we cannot say for sure if they went to the park.

---

4 Doctors say that people.. Answer: B) Bobby will not fall sick.
<u>Reasoning:</u> The conditional statement is:
   people get flu shots → they will not fall sick
Since only Bobby took the flu shot, only she will not fall sick. Choice A is misleading because it suggests that people (all people) will not fall sick because Bobby got the flu shot.

Answers

> INFERENCING-must be true

5 Advertisement: Introducing the new pureWhite soap..
Answer: B) Reasoning: The conditional statement is:
  pureWhite is used in dirty white clothes → become clean in two minutes

Joyce used the pureWhite soap on her muddy white pants and it became clean within two minutes. This must be true as it satisfies the condition that the soap is used on a dirty white cloth (muddy white pant). Choice A is tricky. It also uses the soap in a muddy white shirt, but it says that the shirt became clean within three minutes. If it took two and a half minutes to become clean, it does not satisfy the claim that it will become clean within two minutes. So, it is important to read all the facts carefully.

---

6 Banner in library: "If you read adventure books..
Answer: A) Mary will be thrilled.
Reasoning: The conditional statement is:
   you read adventure books → will be thrilled
Since Mary and John both read two adventure books, both of them will be thrilled. It cannot be true that John was bored.

---

7 Cat to Mouse: If you run away.. Answer: A) The cat ate the mouse.
Reasoning: The conditional statement is:
   you run away → I will not eat you
Since the mouse ran away, it satisfied the condition. So, it must be true the cat did not eat the mouse.

Answers

INFERENCING-must be true

8 Fox to Crow: If you sing a song..
Answer: B) The Fox was happy.
Reasoning: The conditional statement is:
  you sing a song → i will be happy
Since it was the Fox that imposed the condition, it is the fox that will be happy if the crow sings and we know that the crow sang.

---

9 Policeman: If you cross the street..

Reasoning: The conditional statement is:
  cross the street during 'do not walk' → you will get a ticket.
Since Benjamin and Martha both crossed the street when it said 'do not walk', both of them will be fined.

1) Benjamin was not fined.   Answer: B) False
2) Martha was fined.   Answer: A) True

INFERENCING-cannot be true

10 Notice at the botanical gardens: If it is Spring the birds will come!
Answer: A) The birds did not come even though it is Spring.
Reasoning: The conditional statement is:
  Spring → birds will come

Choice A says that the birds did not come even though it is Spring. This clearly violates the conditional statement and so this choice cannot be true. The choice that cannot be true is the correct answer.

Answers

INFERENCING-cannot be true

11 If Gibson comes to the picnic, then Larry will come to the picnic.
Answer: B) Gibson came to the picnic, but Larry did not.
<u>Reasoning:</u> The conditional statement is:
   Gibson comes → Larry will come
Choice B says that Gibson came, but Larry did not. This violates the condition and hence cannot be true.

12 Doctor to Kerry: If you take this medicine, you will get well.
Answer: A) Kerry did not get well in spite of taking the medicine.
<u>Reasoning:</u> The conditional statement is:
   you take this medicine → will get well
Choice A cannot be true as it violates the condition.

13 Bookstore: If you pay for the book today, it will be available the next day. Answer: A) Tom paid on Tuesday and got the book on Thursday.
<u>Reasoning:</u> The conditional statement is:
   pay today → available next day
Choice A cannot be true as it violates the conditional statement. Tom should receive his book on Wednesday.

Answers

INFERENCING-unconditional-must be true

15 Sam hit the ball..

Answer: B) The window is outside the fence.

Reasoning: The ball went outside the fence and smashed the window. So, we can conclude that the window is outside the fence.

16

The water comes out of the shower..

Answer: A) The water comes out of the pipe.

Reasoning: The water comes out of the shower head. But, since the shower head is attached to the pipe, we can conclude that water comes out of the pipe. The water comes out of the pipe and then from the shower head. Choice B is incorrect since the shower head is attached to the pipe, not the wall.

17

Airplanes carry.. Answer: B) Airplanes carry the bags.

Reasoning: Since airplanes carry passengers and passengers in turn carry their bags, it must be true that airplanes carry the bags.

18

The ship is..

Answer: A) Passengers are sailing.

Reasoning: Since the passengers are in the ship and the ship is sailing, it must be true that the passengers are also sailing.

Answers

INFERENCING-unconditional-must be true

19 From wood, we can make paper ..
Answer: B) From wood, we can make notebook.
Reasoning: You can draw a simple arrow diagram to help you visualize the relationship and pick the correct answer.

Wood → Paper → Notebook.

The relation expressed by arrows is "make" - from wood we make paper, from paper we make notebook. So, we can conclude that from wood we can make notebook. From notebook we don't make wood and so, choice A is incorrect.

20 The train is going north. Tracy is sitting in the train facing south.
Answer: B) Tracy is going north.
Reasoning: From common sense, we know that if the train is going north, everything inside will also go north, regardless of which direction it is facing. So, Tracy is also going north.

21 All birds are beautiful. Peacock is a bird.
Answer: B) Peacock is beautiful.
Reasoning: Since all birds are beautiful, and peacock is a bird, it follows that peacock is beautiful.

22 All the students in a class are reading. Nick is a student in this class. Answer: B) Nick is reading.
Reasoning: Since all the students in a class are reading and Nick is a student in this class, he is also reading, not writing. Note carefully the use of the words "all" and "in this class". If Nick was a student of some other class, then we cannot conclude that he is reading.

Answers

© Gift Of Logic, Inc * Copying prohibited

INFERENCING-unconditional-must be true

23 Floor P is below floor Q..
Answer: B) Floor R is below floor Q.
Reasoning: To clear up the confusion with the floors, a simple diagram like the one below will help you figure out where each floor is.

    Q
    P
    R

24 There will be a test on Logic before lunch.
Answer: B) When the lunch begins, the test would have been taken.
Reasoning: Remember the sequence of events - first the test and then lunch. Note how the answer choices are presented in such a way as to confuse you about the correct sequence of events. "When the lunch begins, the test would have been taken" is the same as saying that test happened first and then the lunch, which must be true.

25 During a field trip, Amy's class will visit a farmland..
Answer: A) Amy's class will see tractors before they see rockets.
Reasoning: The word "then" indicates that farmland is visited first and then the Space Center. Note that the answer choices use the word "after" to describe the order of the events. Since they will visit farmland first, they will see tractors first and then rockets.

26 Jim is driving his car at 55 miles per hour..
Answer: B) Only Jim is driving above the speed limit.
Reasoning: Note the words "only" and "both" in the answer choices.

Answers

INFERENCING-unconditional-must be true

27 A rabbit, a fox and a lion are standing in a line ...
You can use letters R for Rabbit, F for fox and L for Lion and then represent their positions as follows:
$$R, F, L$$
Answer: A) The lion is standing behind the rabbit.
Reasoning: If you look at the representation, it is easy to find that this choice must be true.

---

28 A tree has several branches..
Answer: B) A tree has several leaves.
Reasoning: Represent the facts as follows: Tree - Branches - Leaves. "-" here represents the word "has".

---

29 Everyone in Tim's class is..
Answer: B) John is more than three feet tall.
Reasoning: Note the use of the word "exception". This means that John does not satisfy the condition that all students in the class are at most three feet tall. Note the use of the word "at most". This means that no one is taller than three feet except John who is the exception.

---

30 All stores are open this Sunday except the Bond bakery.
Answer: B) Nothing can be purchased at the Bond bakery this Sunday.
Reasoning: Bond bakery is not open this Sunday. So, nothing can be purchased from it this Sunday. Choice A is incorrect because it does not refer to this Sunday, but concludes that nothing can be purchased at the Bond bakery. This is not true.

Answers

INFERENCING-unconditional-cannot be true

31 Two trains are going in the same direction ..

Answer: A) The other train is going south.
Reasoning: Read carefully that the two trains are going in the same direction and one is going north. So, obviously the other train also must be going north. It cannot be going south. The choice that cannot be true is the correct answer.

---

32 The two flowers in the flower vase..

Answer: B) The other flower is red in color.
Reasoning: Read carefully that the two flowers are of different colors and one is red. So, the other cannot be red as well.

---

33 A leader's job is not an easy job. Jack is the leader of his class.

Answer: A) Jack has an easy job.
Reasoning: We can infer from the given statements that Jack's job is a tough job. So, to say that Jack has an easy job cannot be true.

---

34 The group of people in this room are not boys ..

Answer: A) The person wearing a mask is a boy.
Reasoning: Since the person wearing a mask belongs to the group and since the group does not have boys, we can infer that the person wearing a mask is not a boy. Therefore, the statement that the person wearing a mask is a boy cannot be true.

Answers

© Gift Of Logic, Inc * Copying prohibited

INFERENCING-unconditional-cannot be true

35 It does not matter whether you travel in flight# 33 or flight# 43..

Answer: B) Flight# 43 takes fifty minutes, whereas flight# 33 takes forty minutes to reach London.
Reasoning: This cannot be true since the given statements indicate that both flights take the same time to reach London.

---

36 One end of a battery is positive..

Answer: A) Both ends of batteries made by Bright Industries are positive.
Reasoning: This choice cannot be true since it is inconsistent with the stated facts.

---

37 Several students stood in a line...
Answer: B) The fourth student was seen by the doctor before the second student.
Reasoning: This cannot be true since the doctor will only see patients one by one starting from the first student. So, the second student will be seen before the fourth student.

---

38 Jack can lift up to 100 pounds ...

Answer: B) Jack can lift whatever Jim can.
Reasoning: This cannot be true because Jim can lift more than 100 pounds, but Jack can lift only up to 100 pounds. Jack cannot lift whatever Jim can.

Answers

INFERENCING-unconditional-cannot be true

39 Temperatures are warmer in summer ..

Answer: A) Water does not evaporate more during summer.

Reasoning: This cannot be true.

summer → warmer temperatures
warmer months → water evaporates more
so, summer → water evaporates more

The statement expressed by choice A that water does not evaporate more during summer cannot be true.

40

The candy box can be placed inside..

Answer: A) Popcorn box can be placed inside the candy box.

Reasoning:
    Candy box can be placed inside the popcorn box.
Vice-Versa of the above statement is:
    The popcorn box can be placed inside the candy box.

The given statement clearly says "not vice versa". So, the popcorn box cannot be placed inside the candy box. To say that the popcorn box can be placed inside the candy box cannot be a true statement.

Answers

© Gift Of Logic, Inc * Copying prohibited

INFERENCING-unconditional-cannot be true

41 Jack went to Jill's house and vice versa.

Answer: B) Jill did not go to Jack's house.

Reasoning: The vice versa, or converse of "Jack went to Jill's house" is "Jill went to Jack's house". Choice B says that Jill did not go to Jack's house, which cannot be true as it contradicts the given facts.

---

42 Brendan was standing third in line..

Answer: A) Brendan is ahead of Barak in the line.

Reasoning:

Before the swap, the positions of Brendan and Barak were:
    Brendan - 3   Barak - 5.
So, Brendan was head of Barak in the line before the switch.

After the swap, the positions of Brendan and Barak are:
    Brendan -5   Barak - 3
So, Barak is ahead of Brendan in the line after the switch.

So, choice A that says that Brendan is ahead of Barak in the line after the switch cannot be true.

Answers

AGREE

1 Nurse: Eating food ..

Your reply: Eating food with too much sugar can cause the blood pressure to increase above normal levels. This is not good for your health.

---

2 Car driver: While driving a car..

Your reply: Keeping a safe distance from the car in front of us will help us avoid an accident if it stops suddenly.

---

3 Civil Engineer: Bridges must..

Your reply: There will be a lot of vehicles traveling over the bridge. So, the bridge must have a strong foundation to withstand the heavy load.

---

4 Policeman: Wearing a helmet is very important when you ride your bike.

Your reply: The helmet helps to protect us from head injuries. So, it is necessary to wear helmets when we ride our bikes.

Answers

AGREE

5

Fox: The lion .. is majestic. So, it is the king of the jungle.
Pig: The lion is powerful. So, it is the king of the jungle.

Answer: B) that the lion is the king of the jungle. Both of them clearly agree that the lion is the king of the jungle. But, the fox thinks that the lion is majestic and that's the reason why it is the king of the jungle. The pig thinks that the lion is powerful and that's the reason why it is the king of the jungle. So, they both agree that the lion is the king of the jungle for different reasons.

6

Molly: A handbag is very useful..
Dolly: A handbag is useful..

Answer: B) that they like to carry a handbag with them wherever they go. Both of them say that they like to take their handbags with them wherever they go, even though the reason they give for doing so is different.

7

Doctor: Since the baby..
Nurse: The baby has..

What do the Doctor and the Nurse agree about?
Answer: A) that the baby needs medicine. This is clear from the statements of the doctor and the nurse.

AGREE

8

Pilot: This plane can ..
Co-Pilot: This plane has..

1) Do the Pilot and Co-Pilot agree that the plane is very big?
Answer: A) Yes. This is clear from their statements.

2) Do the Pilot and Co-Pilot agree on why they think the plane is very big?
Answer: B) No. They give different reasons as to why they think the plane is very big. The Pilot thinks that the plane is very big because it can seat 500 passengers. The Co-Pilot thinks that the plane is very big because it has very wide wings.

DISAGREE

1

Sarah: We should communicate clearly by talking very loudly.

Your reply: It is not necessary to talk loudly in order to communicate clearly. We can communicate clearly just by talking gently.

2

Sanjay: Reading one story book every day is the best way to improve our vocabulary.
Your reply: We don't need to read a story book to improve our vocabulary. Instead, we can learn a few words from the dictionary everyday.

Answers

DISAGREE

3 Jenny: Dora is boring. So, her birthday party will also be boring.
Your reply: Dora has invited a clown for her birthday party. So, the party will be exciting.

---

4 Justin: There is no use playing Chess.
Your reply: Playing Chess is useful to develop critical thinking skills.

---

5 Bonny: The temperature today is much colder than it was yesterday.
Your reply: It is warmer today compared to yesterday.

---

6 Susan: Riding a bike to school is not good for your health.
Your reply: Riding a bike to school is a good exercise that can improve your health.

---

7 Farooq: Eating french fries is good for health because it is tasty.
Your reply. Just because it is tasty does not mean that it is good for your health. In fact, french fries have a lot of fat in them that is not good for your health.

---

8 Kathy: We should not keep in touch with our old friends.
Your reply: Contrary to what you say, we should keep in touch with our old friends by frequently writing letters or talking to them and learn about the changes in their lives.

Answers

1 LIST PROCESSING   position

A monkey, a squirrel, and a parrot sat in three positions..

Write the names of the occupants in the list below.

Position	1	2	3
Occupant	Monkey	Squirrel	Parrot

The parrot flew away and an eagle took its place instead. Now, modify the list to reflect this change.

Position	1	2	3
Occupant	Monkey	Squirrel	Eagle

The squirrel jumped out of the branch and the monkey moved to take its place. Now, modify the list to reflect the change.

Position	1	2	3
Occupant		Monkey	Eagle

Finally, the monkey and the Eagle interchanged their positions. Modify the list to reflect this change.

Position	1	2	3
Occupant		Eagle	Monkey

Answers

LIST PROCESSING

2 Serena went to the grocery store ..
In the list below, write the items that Serena must buy now.

> peanut butter
> milk
> orange juice
> green peas
> popcorn

3

The flights from London and Miami..

Flight Arrivals - On Time		
Flight#	From	Gate
133	Frankfurt	35
146	Mumbai	17

Flight Arrivals - Delayed		
Flight#	From	Gate
143	London	31
121	Miami	22

Answers

LIST PROCESSING

4 Alex had a list of friends..

Only Ali and Roger are in town and have not been called yet. His friends who are marked with a • cannot be invited since they are not in town.

Ali
Roger

5

Bruce Lee threw a birthday..
1) Which of the following people did Wang not see at the party?

Answer: C) Arthur. Wang did not see Arthur. That's why he did not include Arthur in his list. But, Tang saw Arthur and included him in her list.

2) Which of the following people did Tang not see at the party?
Answer: C) Gonzales. Tang did not see Gonzales. That's why she did not include him in her list, but Wang saw Gonzales and included him in his list.

3) If both Wang and Tang saw everyone who was at the party, how many people were at the party? Answer: 7 There are seven individuals at the party.

Answers

LIST PROCESSING

6

> Lu
> Margie
> Mary
> Popov
> Steve
> Victor
> Vivek
> Walid

Note that Margie should be listed before Mary since her fourth letter (g) appears before Mary's fourth letter (y) in the alphabetic list. Victor should be listed before Vivek because 'c' comes before 'v'.

> Miller   Mike   Steve   Stacy   Chris

Note that Miller will come before Mike in the descending order and Steve will come before Stacy.

---

7 Sort the list by grade

best grade first

> Linda
> Farha
> Jenny
> Darren
> Trevor

worst grade first

> Trevor
> Darren
> Jenny
> Farha
> Linda

Answers

8 LIST PROCESSING

The distance from Dallas..

Houston	300
Waco	100
Richardson	15
Plano	18
Austin	200

Richardson
Plano
Waco
Austin
Houston

1) Which city is the nearest to Dallas?
   Answer: Richardson

2) Which city is the farthest from Dallas?
   Answer: Houston

3) Which city is the second nearest to Dallas?
   Answer: Plano

4) Which city is the third nearest to Dallas?
   Answer: Waco

Answers
© Gift Of Logic, Inc * Copying prohibited

| 1 | **SEQUENCING** |

Sam went to the school..
Rewrite the above sentences in sequence starting from the morning.

Answer: Sam woke up at 6 AM. He went to school at 8 AM. He returned home at 4 PM.

A school hoists a flag..

Rearrange the schedule in sequence starting from the first day of the week.

Monday	Orange
Tuesday	Red
Wednesday	Green
Thursday	Blue
Friday	Yellow

2 SEQUENCING

Write the months of the year from January in sequence.

First month to last month
January
February
March
April
May
June
July
August
September
October
November
December

Now, write the months of the year in reverse sequence starting from December.

Last month to first month
December
November
October
September
August
July
June
May
April
March
February
January

3 SEQUENCING

Alison went for a haircut ..

Answer: B) soccer, haircut, go home, read a book, play in the park, drew pictures, went to bed.

---

beginning/end

Many people were standing in a line to get a free T-shirt.

Who was standing in front of the first person in the line?
    Answer: Nobody. The line begins with the first person.

Who was standing in front of the last person in the line?
    Answer: 9th person.

Ten people were standing..

1) After giving a candy to the fifth person, who will you give the candy to?
Answer: B) Fourth person.

2) After giving candy to the first person, who will you give a candy to?
Answer: B) Nobody. The first person is the last person to give candy to.

Answers
© Gift Of Logic, Inc * Copying prohibited

4   SEQUENCING

Number the boxes shown below sequentially beginning with 1. Also, name the balls between the boxes in alphabetic sequence starting with A.

Answer the following questions based on the sequence of objects that you see in the second figure.

1) Write the number sequence only: Answer: 1,2,3,4,5,6

2) Write the alphabetic sequence only: Answer: A, B, C, D, E

3) Write the mixed sequence: Answer: 1, A ,2, B, 3, C, 4, D, 5, E, 6

4) What are the first and last numbers in the sequence? Answer: 1 and 6

5) What are the first and last alphabets in the sequence? Answer: A and E

6) What number is between alphabets C and D? Answer: 4

7) What alphabet is between numbers 3 and 4? Answer: C

8) What alphabet is after number 6? Answer: Nothing
9) Which alphabet is before alphabet A? Answer: Nothing
10) Which number is after number 6?   Answer: Nothing

Answers

© Gift Of Logic, Inc * Copying prohibited

GROUPING

1

Domestic Animals Group	Wild Animals Group
Cow	Lion
Dog	Fox
Horse	Tiger
Cat	Cheetah

1) How many animals are there in the Domestic Animals group? Ans: 3
2) How many animals are there in the Wild Animals group? Ans: 3
3) To which group will you add a Cat?   Ans: Domestic Animals
4) To which group will you add a Cheetah? Ans: Wild Animals

2

Vegetables	Fruits
Carrot	Grape
Onion	Orange
Cauliflower	Apple

Sort the groups in ascending order.

Vegetables	Fruits
Carrot	Apple
Cauliflower	Grape
Onion	Orange

Note that even though Carrot and Cauliflower both begin with a C, Carrot is listed above Cauliflower in the list because, the third letter of Carrot, namely "r" comes before the third letter of Cauliflower, namely "u".

Answers

| GROUPING | combining |

3

T1	Cold
T2	Headache
T3	Cough
S1	Cough
S2	Headache
S3	Cold

0) Which medicine used for treating headaches is a syrup?
    S2

1) Write the group of medicines used for treating cold.
    T1, S3

2) Write the group of medicines used for treating cough.
    T3, S1

3) Write the group of medicines used for treating cold and cough.
    T1,T3,S1,S3

4) Write the group of medicines used for treating headache and cold.
    T1,T2,S2,S3

Answers

1 VENN DIAGRAM

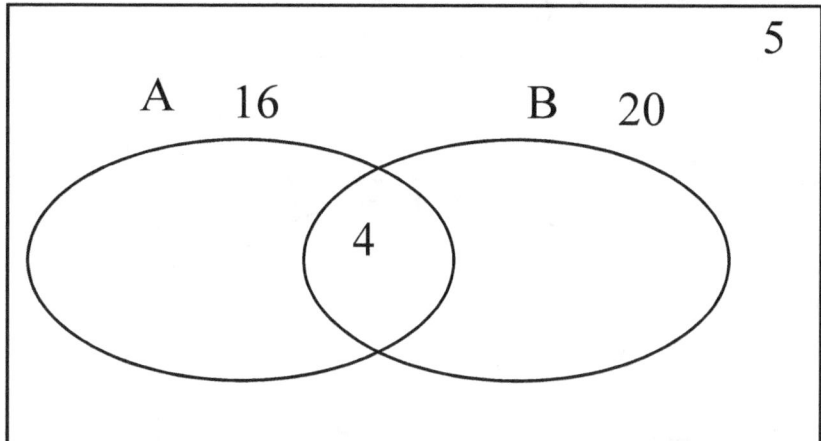

All the people in a room were split into two groups, group A and group B. There are altogether 16 people in group A, and 20 people in group B.
4 people belong to both groups and 5 people do not belong to either group. Based on this information, answer the questions below.

1) How many people are there in the room?
   Answer: (16-4)+ 4+ (20-4) + 5 = 37 people

2) How many people are there in either group?
   Answer: (16-4) + 4 + (20-4) = 32 people

3) How many people belong to group A only?
   Answer: 16-4=12 people

4) How many people belong to group B only?
   Answer: 20-4=16 people

Answers
© Gift Of Logic, Inc * Copying prohibited

2 VENN DIAGRAM

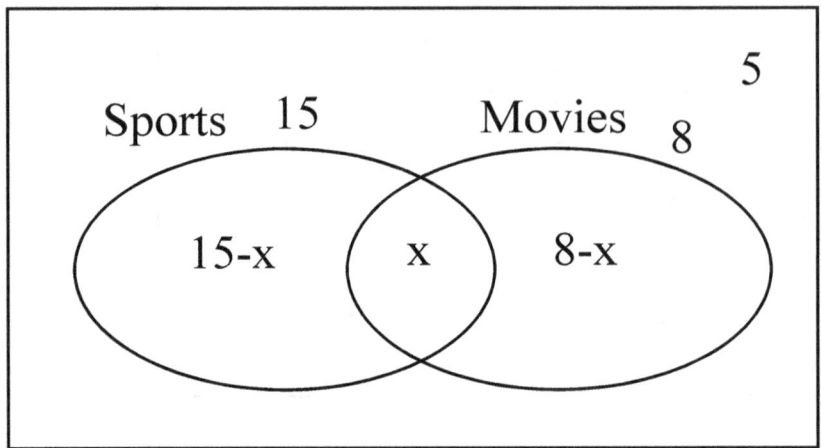

25 students were surveyed and asked if they liked sports or movies. 15 said that they liked sports, 8 said that they liked movies and 5 said that they liked neither. Complete the Venn diagram and answer the following questions.

Complete the Venn Diagram as shown. x indicates students that liked both sports and movies.

$$15-x+x+8-x+5=25$$
$$28-x=25$$
So, $x=3$

1) How many students liked sports as well as movies?
   $x=3$ students

2) How many students liked sports only?
   $15-x=15-3=12$ students

3) Ho many students liked movies only?
   Answer: $8-x = 8-3 = 5$ students

Answers

3 VENN DIAGRAM

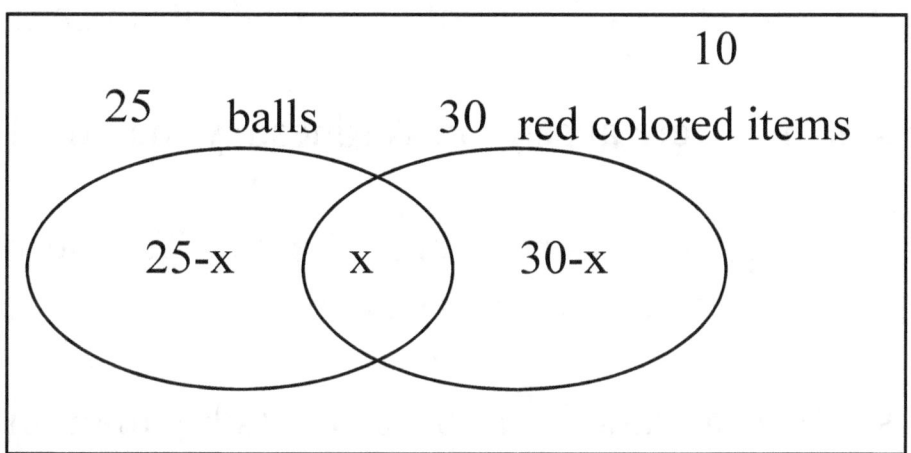

x in the Venn diagram indicates red colored balls.
  25-x+x+30-x+10=50
  65-x=50   so, x=15

1) How many items in the box are either balls or red colored?
   Answer: 25-x+x+30-x= 55-x=55-15=40 items

2) How many items are red colored and balls?
   Answer: x= 15 items

3) How many items are not balls?
   Answer: 30-x+ 10 = 30-15+10= 25 items

4) How many items are not red items?
   Answer: 25-x+ 10 = 25-15+10= 20 items

5) How many items are neither red nor balls?
Answer: 10. Note that this information is given in the problem itself.

Answers

| 1 | GRAPH LOGIC |

The graph above shows the number of students absent..

1) More students were absent on Wednesday than on Tuesday.

Answer: A) True. 10 students were absent on Wednesday whereas only 5 students were absent on Tuesday.

2) More students attended school on Thursday than on Tuesday.

Answer: B) False. This is a tricky question. More students were absent on Thursday than on Tuesday. So, more students attended school Tuesday than on Thursday.

3) From Monday to Thursday, the number of students attending school increased.

Answer: B) False. From the graph, it is clear that the number of students in absence increased from Monday to Thursday. So, the number of students attending school decreased from Monday to Thursday.

Answers
© Gift Of Logic, Inc * Copying prohibited

| **2** | **GRAPH LOGIC** |

The graph above shows the height in feet..

1) Who is the tallest in Paul's family? Answer: B) Dad

2) Who is the shortest in Paul's family? Answer: C) Paul

3) If Paul was three feet taller than what he is now, he will be taller than his dad. Answer: B) False. Paul is three feet tall now. If he is six feet tall, he will be only as tall as his dad, but not taller than his dad.

4) If Paul was three feet taller than what he is now, he will be taller than his mom. Answer: A) True. If Paul was three feet taller, he would be 6 feet tall and therefore, taller than his mom who is only 5 feet tall.

| **3** | **GRAPH LOGIC** |

The graph above shows the Lion and Tiger population..

1) There were more Lions last year than there were Tigers.
Answer: B) False. There were 12 Tigers, but only 10 Lions last year.

2) There are more Tigers this year than Lions.
Answer: B) False. There are 12 Lions, but only 10 Tigers this year.

3) There were the same number of Lions last year as there are Tigers this year. Answer: A) True. There were 10 Lions last year and there are 10 Tigers this year.

Answers

4 GRAPH LOGIC

The weather for one full week..

1) The temperatures increased from 40 degrees to 75 degrees from Sunday to Thursday.

Answer: B) False. The temperature actually fell from 75 degrees to 40 degrees from Sunday to Thursday.

2) Temperatures remained constant for three days in a row.

Answer: B) False. If temperatures remained constant, we would see bars of the same height, but we don't.

3) Temperatures fell more from Sunday to Tuesday than from Tuesday to Thursday.

Answer: A) True. The temperature fell 20 degrees from Sunday to Tuesday and 10 degrees from Tuesday to Thursday.

5 GRAPH LOGIC

Three parks, A, B, and C ..

Read the chart carefully and answer the questions below.

1) Park-A had the same number of visitors as Park-B and Park-C for spring and summer.

Answer: B) False. Park-A had the same number of visitors as park-C in spring and summer, but had different number of visitors than park B in spring.

2) Park-A had the same number of visitors as Park-C in summer and winter.

Answer: A) True. Park had and Park C both had 30 and 40 visitors respectively in summer and winter.

Answers
© Gift Of Logic, Inc * Copying prohibited

| **6** | **GRAPH LOGIC** |

Jack, Jill, and Jane shared..

1) Jack and Jill ate the same amount of pizza.
Answer: B) False. Jack ate more than Jill. Jack ate 60% of the available pizza whereas Jill ate 20% of the available pizza.

2) Jane and Jill ate the same amount of pizza.
Answer: A) True. Both ate 20% of the available pizza.

3) Jane and Jill ate the same number of pizza slices.
Answer: C) Uncertain. Both ate 20% of the available pizza, but information is not available to say for sure that they ate the same number of pizza slices. If the slices are of different sizes, then they could have eaten the same amount of pizza with different number of slices.

| **7** | **GRAPH LOGIC** |

The graph above shows the scores ..

1) Half the number of students scored more than 60 points.
Answer: B) False. Out of 6 students, only one scored more than 60 points.

2) One third of the students have the lowest score.
Answer: A) True. Two students scored 50 points, which is the lowest score shown in the histogram. These two students make up one third of the six students.

NUMBER LOGIC

Figure out the logic in the number sequence and write the missing number.

1 1   3   5   ?   Answer: 7. Numbers increase by 2.

2 2   5   8   ?   Answer: 11. Numbers increase by 3.

3 11   22   33   ?   Answer: 44. Numbers increase by 11.

4 10   20   ?   40   Answer: 30. Numbers increase by 10.

5 ?   7   6   5   Answer: 8. Numbers decrease by 1

6 10   100   1000   ?   Answer: 10000. Numbers multiply by 10.

7 10000   1000   ?   10   Answer: 100. Each number is the previous number divided by 10.

8 0.5   1.5   2.5   ?   Answer: 3.5. Add 1 to the previous number to get the next number.

9 9.5   7   ?   Answer: 4.5. Subtract 2.5 to get the next number.

10 0.25   0.75   2.25   ?   Answer: 6.75. Multiply the previous number by 3 to get the next number.

Answers
© Gift Of Logic, Inc * Copying prohibited

NUMBER LOGIC

Figure out the logic in the numbers and find the missing number.

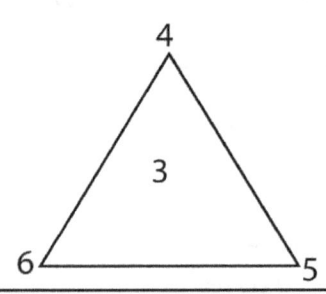 **11**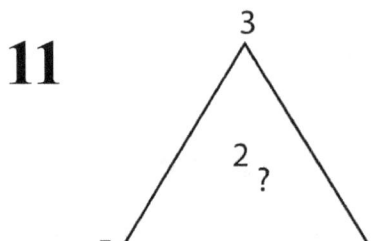

Logic: The numbers decrease anti-clockwise- 6, 5,6 and 3 on left. Similarly, 5,4,3 and 2 on right.

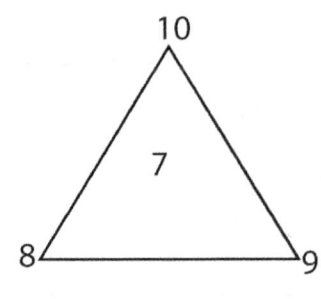 **12**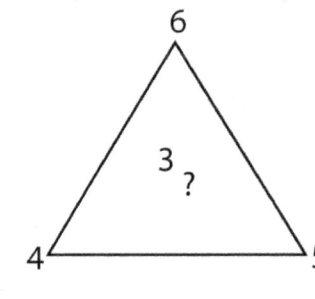

Logic: The numbers decrease clockwise, 10, 9, 8, 7 on left. Similarly, 6,5,4 and 3 on right.

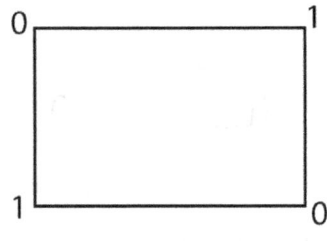 **13**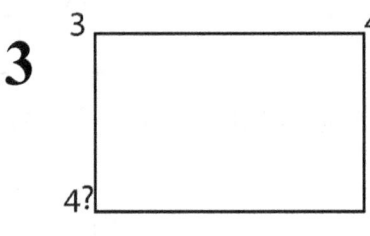

Logic: The numbers in the diagonal are same-0,0 and 1, 1 on left. Similarly, 3,3 and 4,4 on right.

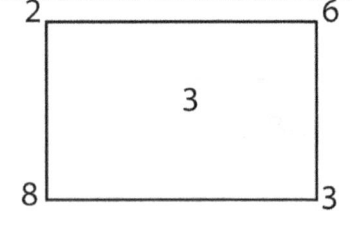 **14**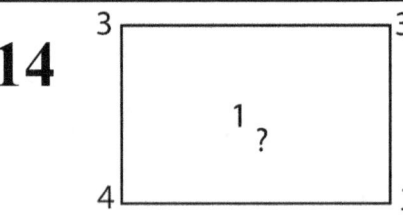

Logic: On the left, add the difference of the numbers in the diagonal. 8-6 + 3-2 = 3. Similarly, on right, 4-3 + 3-3 =1

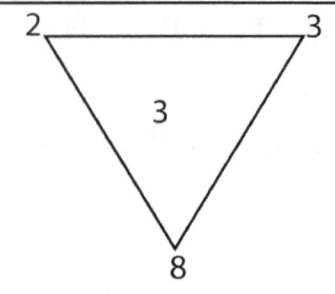 **15**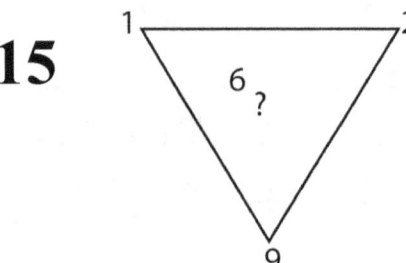

Logic: On the left, 8-3-2= 3, the number inside. On right, 9-2-1=6

Answers
© Gift Of Logic, Inc * Copying prohibited

NUMBER LOGIC

Figure out the logic in the numbers and find the missing number.

16 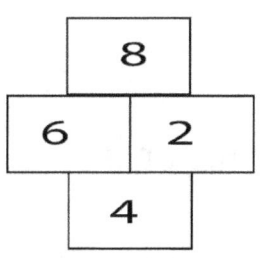 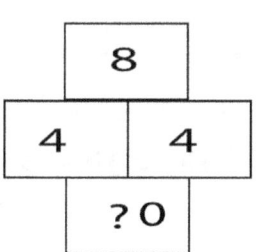 | Logic:. On the left, 6+2=8 and 6-2=4. On the right, 4+4=8, 4-4=0.

17 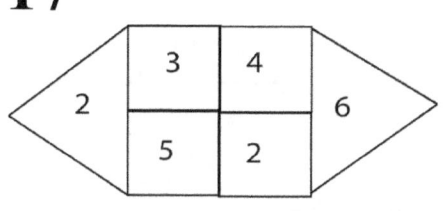 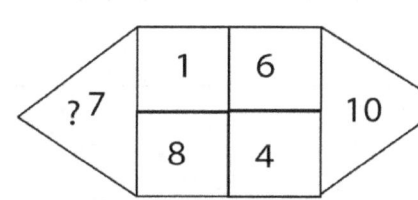 | Logic: On the left, 5-3=2 and 2+4=6. On the right, 8-1=7 and 4+6=10.

18 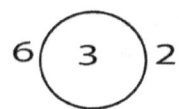  | Logic:. On the left, 2 x 3 =6, on the right, 4 x 2 =8.

19 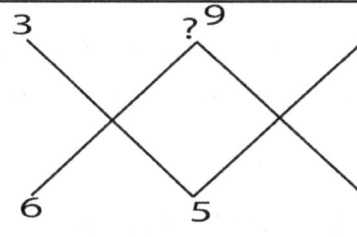 | Logic:. The numbers in the end add up to make the number on the vertex. On the left 1+2=3 and 3+ 4= 7. On the right, 6+ 3=9.

20 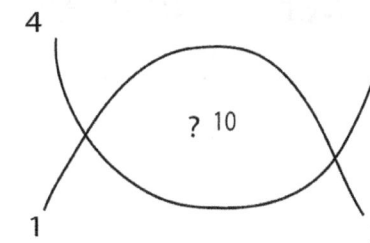 | Logic:. The number at the ends add up to the number in the middle. 4+2+3+1=10.

Answers
© Gift Of Logic, Inc * Copying prohibited

LETTER LOGIC

Figure out the logic in the sequence and find the missing letter or number.

1

A   C   E   ?   Answer: G. Sequence skips one letter in between.

2

K   ?   Q   T   Answer: N. Sequence skips two letters in between.

3

Z   Y   X   ?

Answer: W. Sequence is reverse of alphabets.

4

Z   X   V   ?

Answer: T. Sequence skips one alphabet in reverse.

5   A   1   B   2   C   ?

Answer: 3. Sequence of alphabets and numbers.

6   AB   CD   ?   GH

Answer: EF.   Two alphabets strung together in sequence.

7   1 A 2 B ? C   Answer: 3. Numbers and alphabets in sequence.

8

Z1   Y2   ?   W4

Answer: X3. Numbers increase in sequence while the alphabets progress in reverse order.

Answers   150

© Gift Of Logic, Inc * Copying prohibited

LETTER LOGIC

Figure out the logic in the sequence and find the missing letter or number.

9

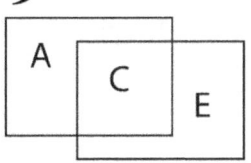 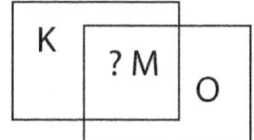

Logic: On the left, A,C,E have one skipped between them. On the right, K first and then skip one to get M

10

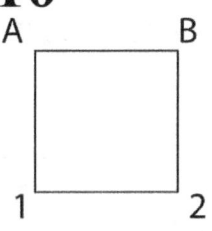 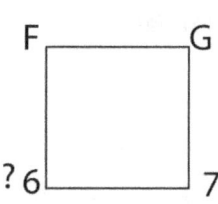

Logic: On the left, A and B correspond to 1 and 2, their position in the alphabet sequence. On the right, G is 7 and F is 6 in the sequence.

11

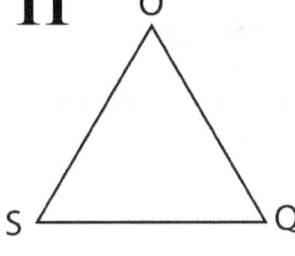 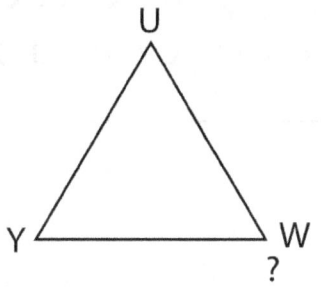

Logic: On the left, O, Q and S are alphabets that skip one between them. On the right, U, W and Y skip one between them.

12

 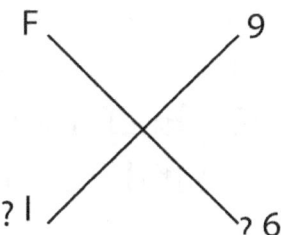

Logic: On the left, C is the 3rd and Y is the 25th in the alphabetic sequence. On the right, F is the 6th and I is the 9th.

Answers

LETTER LOGIC

Figure out the logic in the sequence and find the missing letters.

13

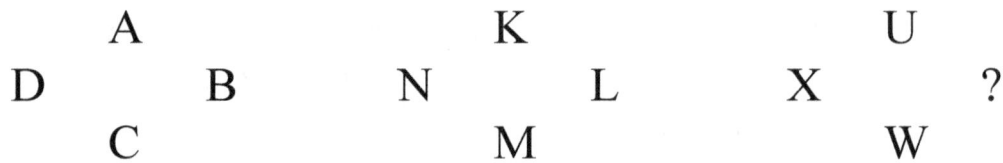

Answer: V. The alphabets are positioned in a circular pattern, listed clockwise.

14

AB YZ CD ? Answer: WX. Two letters from the beginning and two letters from the end alternate each other in sequence.

15

AC BD CE ?   Answer: DF. The first letters are sequential starting from A. The second letters are sequential starting with C.

16

BA DC FE ? Answer: HG. Each member of AB, CD, EF, GH is reversed. Another reasoning is that the first letter of each member is in sequence after skipping one letter. B,D,F and then H; Similarly, for the second letter in each member, the sequence is A, C, E and then skip one to get G.

Answers

© Gift Of Logic, Inc * Copying prohibited

1 SUDOKU

Solve the following Sudoku. A correctly solved Sudoku has numbers 1-9 appearing only once in each row, each column and each 3x3 grid. You gain valuable positioning skills by solving these sudokus.

1	5	3	9	7	8	4	2	6
7	8	4	2	3	6	9	1	5
2	6	9	5	4	1	7	3	8
5	4	1	7	6	3	2	8	9
8	3	6	4	2	9	1	5	7
9	2	7	8	1	5	3	6	4
3	7	5	1	8	4	6	9	2
6	9	2	3	5	7	8	4	1
4	1	8	6	9	2	5	7	3

Answers

2 SUDOKU

Solve the following Sudoku. A correctly solved Sudoku has numbers 1-9 appearing only once in each row, each column and each 3x3 grid. You gain valuable positioning skills by solving these sudokus.

4	3	1	6	7	9	8	5	2
2	6	5	8	1	3	9	4	7
8	9	7	4	5	2	3	1	6
7	4	2	5	9	1	6	3	8
3	8	6	2	4	7	5	9	1
1	5	9	3	6	8	2	7	4
9	2	8	1	3	4	7	6	5
6	7	4	9	8	5	1	2	3
5	1	3	7	2	6	4	8	9

Answers

3 SUDOKU

Solve the following Sudoku. A correctly solved Sudoku has numbers 1-9 appearing only once in each row, each column and each 3x3 grid. You gain valuable positioning skills by solving these sudokus.

3	7	4	8	6	9	2	5	1
2	9	1	3	7	5	4	8	6
5	8	6	1	4	2	3	9	7
9	1	3	4	8	7	5	6	2
6	4	8	5	2	1	7	3	9
7	5	2	9	3	6	8	1	4
1	3	7	2	9	8	6	4	5
8	2	9	6	5	4	1	7	3
4	6	5	7	1	3	9	2	8

Answers 155
© Gift Of Logic, Inc * Copying prohibited

4 SUDOKU

Solve the following Sudoku. A correctly solved Sudoku has numbers 1-9 appearing only once in each row, each column and each 3x3 grid. You gain valuable positioning skills by solving these sudokus.

2	7	1	5	6	9	4	3	8
8	3	4	2	7	1	9	5	6
5	6	9	3	8	4	2	7	1
9	5	3	8	4	7	6	1	2
7	8	2	1	5	6	3	4	9
4	1	6	9	2	3	5	8	7
3	4	7	6	9	8	1	2	5
1	9	5	7	3	2	8	6	4
6	2	8	4	1	5	7	9	3

Answers

5 SUDOKU

Solve the following Sudoku. A correctly solved Sudoku has numbers 1-9 appearing only once in each row, each column and each 3x3 grid. You gain valuable positioning skills by solving these sudokus.

4	1	7	5	6	2	8	3	9
9	2	6	4	3	8	7	1	5
8	3	5	7	9	1	4	2	6
5	6	8	3	1	9	2	7	4
2	4	1	8	7	6	9	5	3
7	9	3	2	5	4	1	6	8
6	7	4	1	8	5	3	9	2
3	8	9	6	2	7	5	4	1
1	5	2	9	4	3	6	8	7

Answers

PICTURE SEQUENCE

Figure out the logic in the picture sequence and draw the next picture in the sequence.

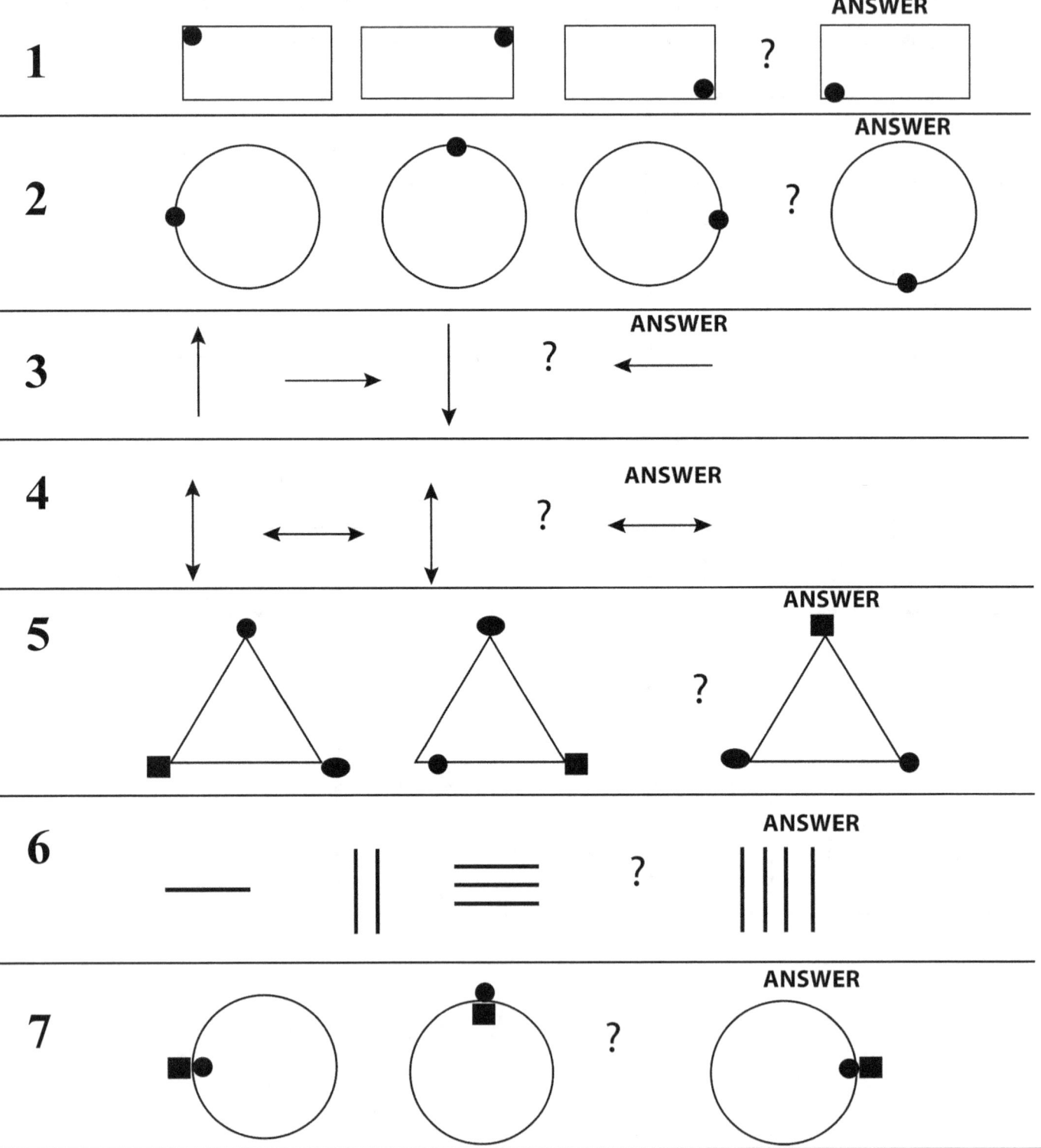

Answers
© Gift Of Logic, Inc * Copying prohibited

PICTURE SEQUENCE

Figure out the logic in the picture sequence and draw the next picture in the sequence.

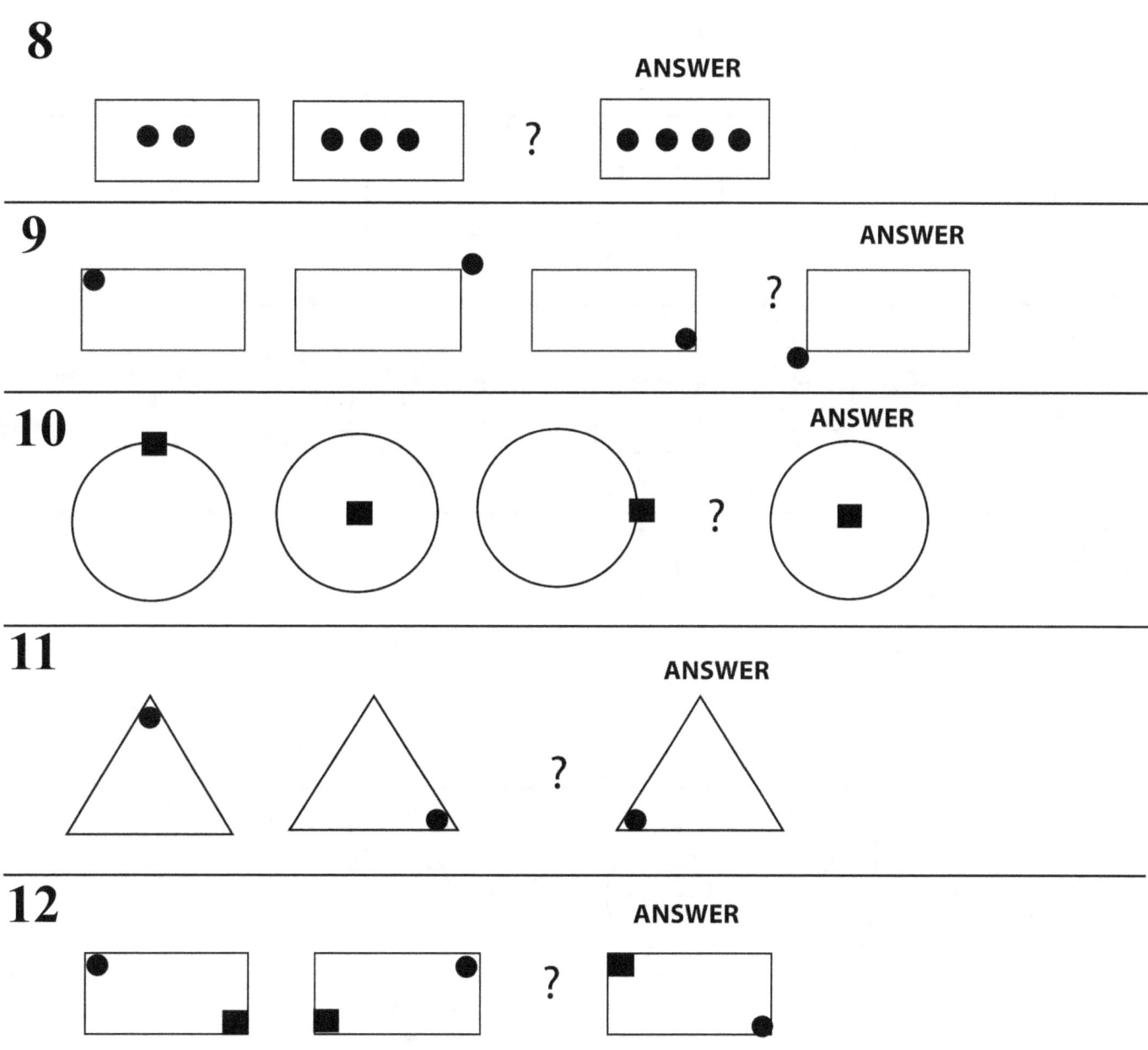

PICTURE SEQUENCE

Figure out the logic in the picture sequence and draw the next picture in the sequence.

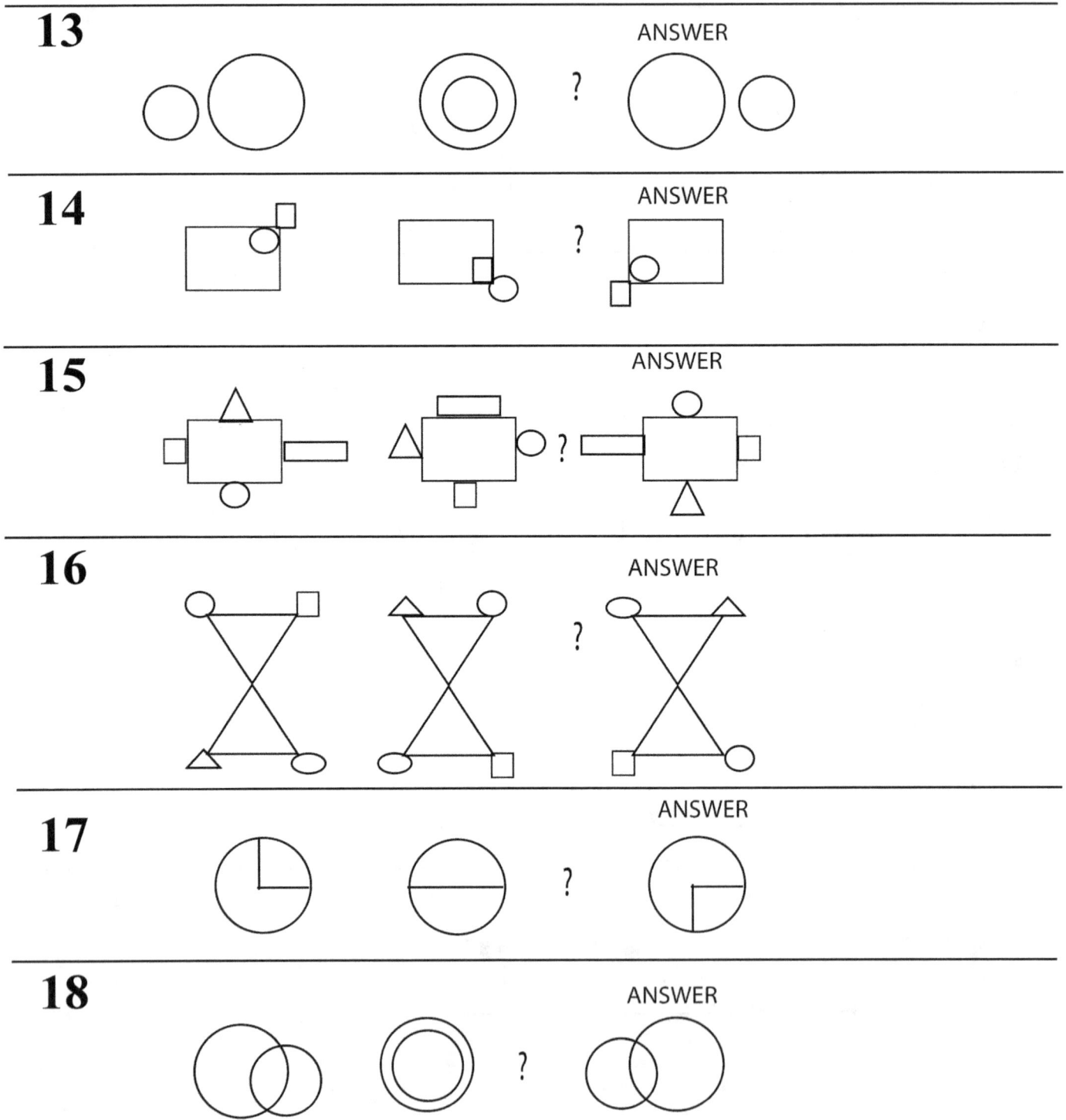

Answers

© Gift Of Logic, Inc * Copying prohibited

PICTURE ANALOGY

Figure out the logic in the picture analogy and circle the correct picture that will complete the analogy.

1

▭ : ▫ AS ⬭ : △(A) ◯(B, ANSWER) ▱(C)

2

(puzzle piece with tab on top) : (puzzle piece with tab on bottom) AS (trapezoid bell, tab up) : (A) (B) (C, ANSWER)

3

| : + AS \ : ⟋(A, ANSWER) ✕(B) ⟍(C)

Answers   161
© Gift Of Logic, Inc * Copying prohibited

PICTURE ANALOGY

Figure out the logic in the picture analogy and circle the correct picture that will complete the analogy..

4

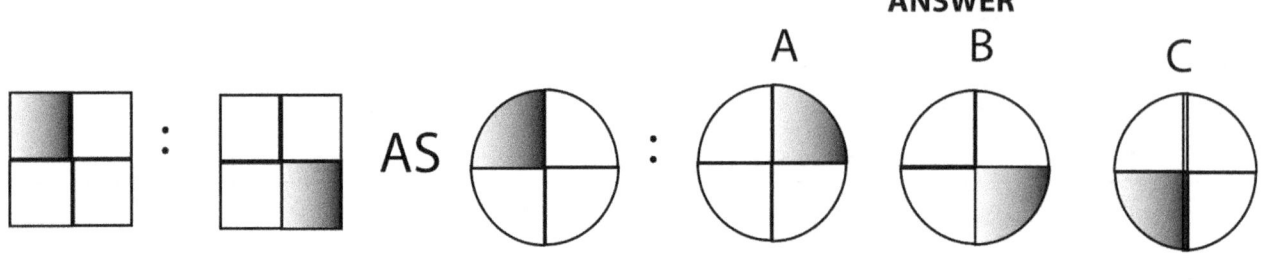

5

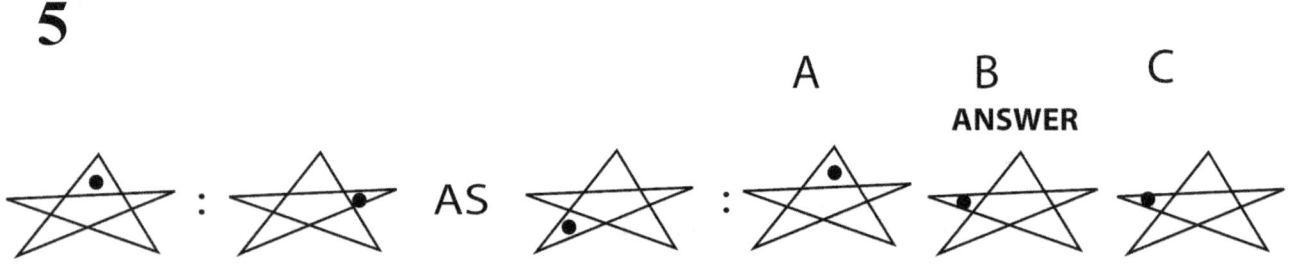

6

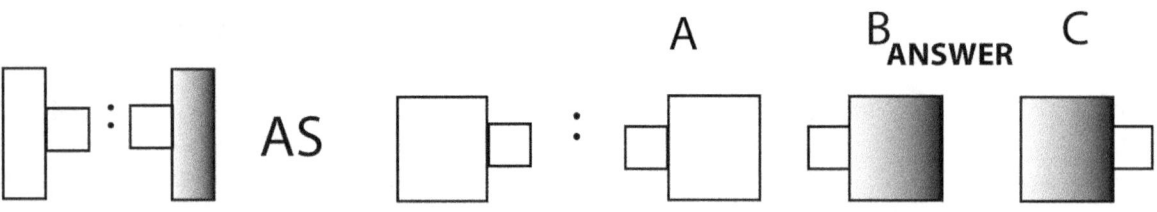

Answers
© Gift Of Logic, Inc * Copying prohibited

PICTURE ANALOGY

Figure out the logic in the picture analogy and draw the picture that will complete the analogy.

7

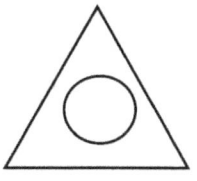 : 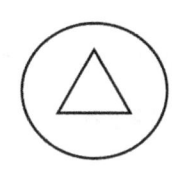 AS 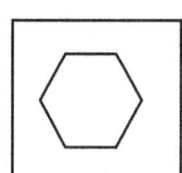 : ?   ANSWER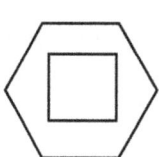

8

 :  AS  : ?   ANSWER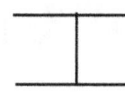

9

 :  AS 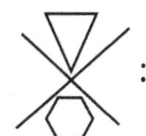  ?    ANSWER

10

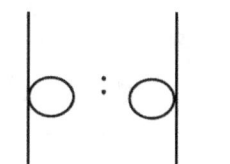 AS 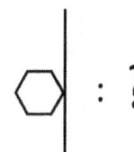 : ?   ANSWER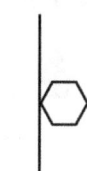

Answers

ODD PICTURE

Find the odd picture in each question below.

1   A   B   C **ANSWER**

 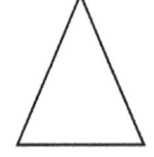 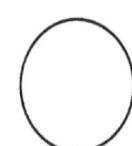 no straight edges in C

---

2   A   B   C **ANSWER**

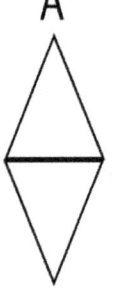   three different shapes in C

---

3   A   B **ANSWER**   C

  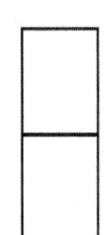 shapes are inside each other in B

---

4   A   B **ANSWER**   C   4-3-1-1 pattern in B

```
V V V V     U U U U     T T T T
V V         U U U         T T
V V          U            T T
V            U             T
```

---

5   A **ANSWER**   B   C

 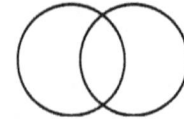 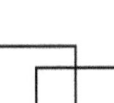 shapes are inside each other in A

Answers

ODD PICTURE

Find the odd picture in each question below.

6 A B C ANSWER

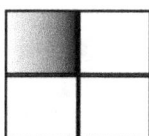 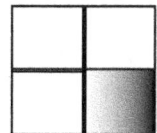   right side is shaded in C

---

7 A B ANSWER C

    all shapes are different in B

---

8 A B C ANSWER

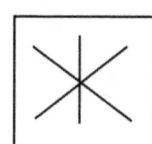 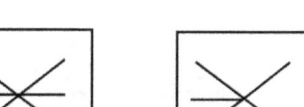   one line is missing in C

---

9 A B ANSWER C

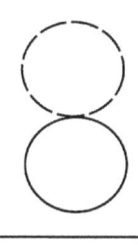 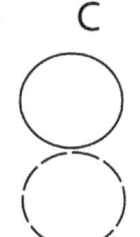   no dotted circle in B

---

10 A B C D ANSWER

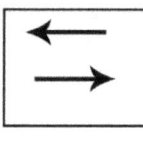 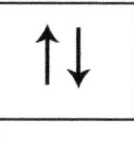 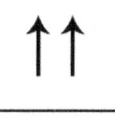   arrows are pointed in same direction in D

Answers

© Gift Of Logic, Inc * Copying prohibited

PICTURE DIFFERENCE

Mark the differences in the set of pictures shown, with arrows.

1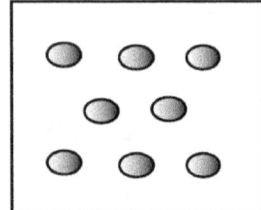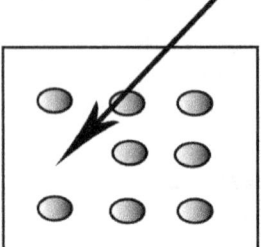

2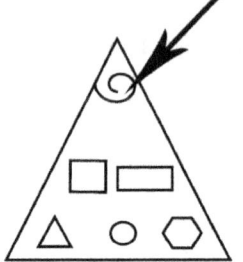

3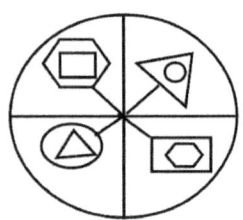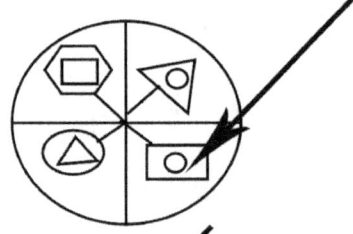

4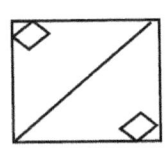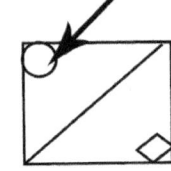

5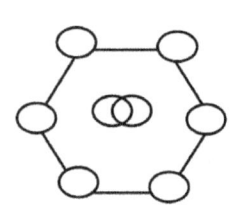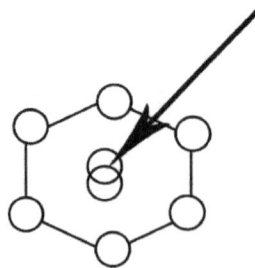

Answers

PICTURE DIFFERENCE

Mark the differences in the set of pictures shown, with arrows.

6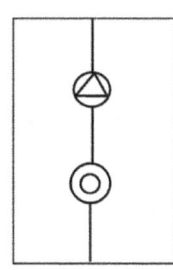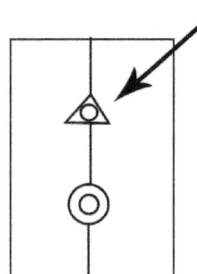

7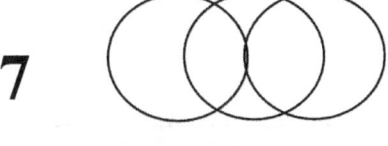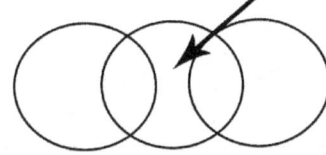

8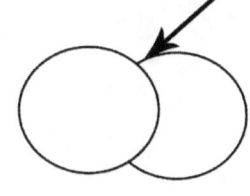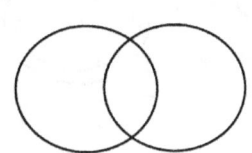

9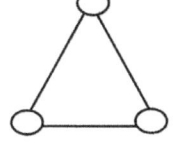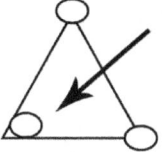

10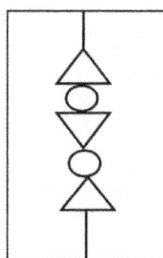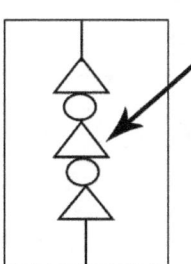

PICTURE DIFFERENCE

Mark the differences in the set of pictures shown, with arrows.

11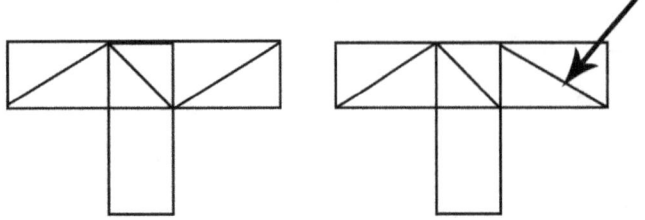

12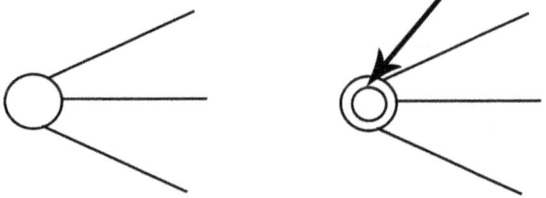

13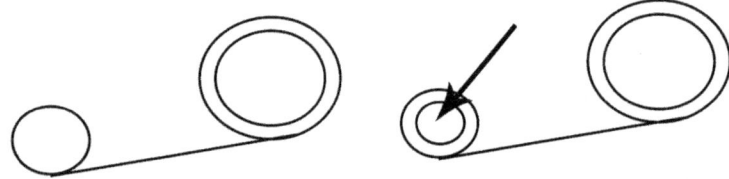

14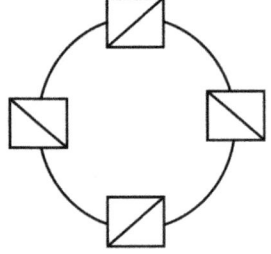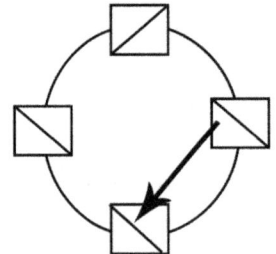

15

Answers 168
© Gift Of Logic, Inc * Copying prohibited

PATTERN MATCHING

Find the pattern in the picture-set on the left and identify the picture on the right that will fit in the space marked with ? to complete the pattern.

1    A   B   C ANSWER
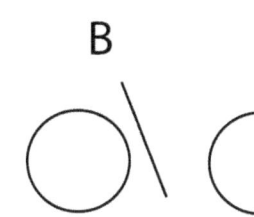

2 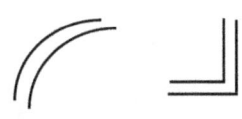   A   B ANSWER   C

3    A   B ANSWER   C

4    A   B ANSWER   C
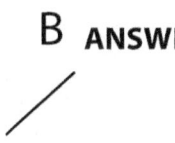

5 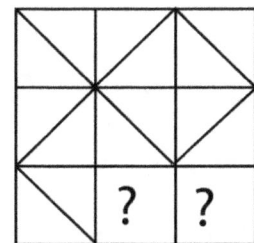   A   B ANSWER   C

Answers   169
© Gift Of Logic, Inc * Copying prohibited

NOTES

NOTES

www.ingramcontent.com/pod-product-compliance
Lightning Source LLC
Chambersburg PA
CBHW080247180526
45167CB00006B/2450